ATMOSPHERE

A Scientific History of Air, Weather, and Climate

DISCOVERING
the
EARTH

ATMOSPHERE

A Scientific History
of Air, Weather, and Climate

Michael Allaby
Illustrations by Richard Garratt

Facts On File
An imprint of Infobase Publishing

ATMOSPHERE: A Scientific History of Air, Weather, and Climate

Facts On File, Inc.
An imprint of Infobase Publishing
132 West 31st Street
New York NY 10001

Library of Congress Cataloging-in-Publication Data
Allaby, Michael.
  Atmosphere: a scientific history of air, weather, and climate / Michael Allaby; illustrations by Richard Garratt.
      p. cm.—(Discovering the Earth)
  Includes bibliographical references and index.
  ISBN-13: 978-0-8160-6098-6
  ISBN-10: 0-8160-6098-3
  1. Atmosphere—Popular works. 2. Climatology—Popular works. 3. Climate—Popular works.
I. Title.
QC863.4.A425 2009
551.5—dc22   2008019503

Facts On File books are available at special discounts when purchased in bulk quantities for businesses, associations, institutions, or sales promotions. Please call our Special Sales Department in New York at (212) 967-8800 or (800) 322-8755.

You can find Facts On File on the World Wide Web at http://www.factsonfile.com

Text design by Annie O'Donnell
Illustrations by Richard Garratt
Photo research by Tobi Zausner, Ph.D.

Printed in China

CP Hermitage 10 9 8 7 6 5 4 3 2 1

This book is printed on acid-free paper.

CONTENTS

PREFACE

Almost every day there are new stories about threats to the natural environment or actual damage to it, or about measures that have been taken to protect it. The news is not always bad. Areas of land are set aside for wildlife. New forests are planted. Steps are taken to reduce the pollution of air and water.

Behind all of these news stories are the scientists working to understand more about the natural world and through that understanding to protect it from avoidable harm. The scientists include botanists, zoologists, ecologists, geologists, volcanologists, seismologists, geomorphologists, meteorologists, climatologists, oceanographers, and many more. In their different ways all of them are environmental scientists.

The work of environmental scientists informs policy as well as providing news stories. There are bodies of local, national, and international legislation aimed at protecting the environment and agencies charged with developing and implementing that legislation. Environmental laws and regulations cover every activity that might affect the environment. Consequently every company and every citizen needs to be aware of those rules that affect them.

There are very many books about the environment, environmental protection, and environmental science. Discovering the Earth is different—it is a multivolume set for high school students that tells the stories of how scientists arrived at their present level of understanding. In doing so, this set provides a background, a historical context, to the news reports. Inevitably the stories that the books tell are incomplete. It would be impossible to trace all of the events in the history of each branch of the environmental sciences and recount the lives of all the individual scientists who contributed to them. Instead the books provide a series of snapshots in the form of brief accounts of particular discoveries and of the people who made them. These stories explain the problem that had to be solved, the way it was approached, and, in some cases, the dead ends into which scientists were drawn.

There are seven books in the set that deal with the following topics:

- Earth sciences,
- atmosphere,
- oceans,
- ecology,
- animals,
- plants, and
- exploration.

These topics will be of interest to students of environmental studies, ecology, biology, geography, and geology. Students of the humanities may also enjoy them for the light they shed on the way the scientific aspect of Western culture has developed. The language is not technical, and the text demands no mathematical knowledge. Sidebars are used where necessary to explain a particular concept without interrupting the story. The books are suitable for all high school ages and above, and for people of all ages, students or not, who are interested in how scientists acquired their knowledge of the world about us—how they discovered the Earth.

Research scientists explore the unknown, so their work is like a voyage of discovery, an adventure with an uncertain outcome. The curiosity that drives scientists, the yearning for answers, for explanations of the world about us, is part of what we are. It is what makes us human.

This set will enrich the studies of the high school students for whom the books have been written. The Discovering the Earth series will help science students understand where and when ideas originate in ways that will add depth to their work, and for humanities students it will illuminate certain corners of history and culture they might otherwise overlook. These are worthy objectives, and the books have yet another: They aim to tell entertaining stories about real people and events.

—Michael Allaby
www.michaelallaby.com

ACKNOWLEDGMENTS

My colleague and friend Richard Garratt drew all of the diagrams and maps in the Discovering the Earth set. As always, Richard has transformed my very rough sketches into finished artwork of the highest quality, and I am very grateful to him.

When I first planned these books I prepared for each of them a "shopping list" of photographs I thought would illustrate them. Those lists were passed to another colleague and friend, Tobi Zausner, who found exactly the pictures I felt the books needed. Her hard work, enthusiasm, and understanding of what I was trying to do have enlivened and greatly improved all of the books. Again, I am deeply grateful.

Finally, I wish to thank my friends at Facts On File, who have read my text carefully and helped me improve it. I am especially grateful for the patience, good humor, and encouragement of my editor, Frank K. Darmstadt, who unfailingly conceals his exasperation when I am late, laughs at my jokes, and barely flinches when I announce I'm off on vacation. At the very start, Frank agreed this set of books would be useful. Without him they would not exist at all.

INTRODUCTION

Since people first learned to cultivate plants and raise domesticated animals, they have been at the mercy of the weather. A single hailstorm can destroy a crop. A drought can cause a famine that perpetuates itself as livestock die and starving people eat their crop seeds. Alternatively, enough rain at the right time and warm sunshine to ripen plants mean food will be abundant. There will be celebrations, with singing and dancing, and people will face the winter with confidence.

Weather matters. Even today, when we know so much about what causes the weather, we cannot control it. Harvests can still fail, and in the poorer countries of the world failure means hunger. Because it is a matter of life and death, people have been trying to understand the weather probably since long before they learned to write down their thoughts and dreams. For most of that time the behavior of the atmosphere was attributed to the whims of supernatural beings, who could be appeased and appealed to, but whose ill temper brought suffering and death. Eventually, though, another idea began to gain ground. Rather more than 2,000 years ago in the Greek communities of the eastern Mediterranean, philosophers realized that weather phenomena result from natural causes. It is not the gods that bring the weather, good or bad, but entirely natural processes that men and women might, perhaps, learn to comprehend. Thus was born the scientific study of the atmosphere.

Atmosphere, one of the seven volumes in the Discovering the Earth set, tells the story of the atmospheric sciences. The book begins with the recognition that air is a material substance, a mixture of gases, and describes the unraveling of its chemical composition. The volume goes on to tell of the invention of the barometer and thermometer, which are the most basic of meteorological instruments, and how they came to be calibrated, principally, but not only, by Daniel Fahrenheit and Anders Celsius.

Weather consists mainly of water in one or another of its forms, and the third chapter describes the investigation of clouds and the way they develop and the origin of the names they bear. Air temperature and pressure vary from place to place, from time to time, and, most important, with elevation. The fourth chapter tells of the discovery of the relationships between temperature, pressure, and height above sea level. Air also moves. Winds are very variable in temperate latitudes, but in the Tropics the trade winds are the most dependable winds in the world. This intrigued scientists, whose explanations for why this is led to a wider explanation of the way air transports heat away from the equator. This chapter also recounts the origin of the world's most common system for classifying winds.

Despite air being everywhere around us, until late in the 18th century the atmosphere was largely inaccessible. However, as soon as balloons began to ascend into the sky, meteorologists began to clamber on board clutching their instruments, which is the subject of chapter 6. As information accumulated from studies of the upper air, a wider picture of the atmosphere began to emerge, revealing its structure from the surface all the way to the edge of space. The story then advances to the late 19th and early 20th centuries and the construction of the theory of air masses and frontal systems that underpins modern meteorology. Chapter 8 describes how climates came to be classified.

The realization that weather results from natural causes raised the possibility of predicting it. Weather mapping and forecasting are the subjects of chapter 9, which ends with the discovery of what may prove to be an absolute limit that makes long-range forecasting no better than guesswork. Finally, the book ends with the recognition that climates are constantly changing and that sometimes the changes are dramatic.

What Is Air?

Air is a substance, a mixture of gases, together with droplets of water, particles of dust, crystals of salt and sulfate, spores from fungi and bacteria, and other tiny fragments of material blown up from Earth's surface. Water droplets form clouds, but where there are no clouds the sky is blue.

It all seems very obvious, common knowledge that everyone possesses. And so it is, but only up to a point, because air is not quite like other substances. The atoms, molecules, and most of the particles that make up the air are far too small to be seen by the naked eye. Atoms and molecules of atmospheric gases are too small to be visible even to the most powerful electron microscope. So air is invisible. It is also odorless and tasteless. If it has a smell, the smell is that of some polluting substance, not of the air itself. If it makes a sound, the sound is actually made by more substantial objects or substances. The wind may howl through the telegraph wires, but it is the wires that vibrate to make the sound. Wind turns the sails of windmills and of wind turbines generating power. It drives sailboats and the majestic tall ships that grace the oceans. But what is wind made of? Is it made of anything at all, or is it a force, like gravity? Hold a ball at arm's length and release it and the ball moves downward, never upward. It is drawn toward the Earth by the force of gravity, but no one supposes that gravity is made of any material substance. You cannot bottle gravity. So why should the wind not be a similar force, able to exert pressure but not made of some material that can be contained and moved around?

The realization that substances can exist as invisible, odorless, tasteless gases developed in the 17th century. In the centuries that followed, little by little scientists discovered the composition of air and the properties of its constituent gases. Their search was motivated by intellectual curiosity, but it was curiosity with very practical relevance, because whatever air might be, it is the source of the weather. Lives depended on good harvests and good harvests depended on the weather. Farmers needed to know when it was safe to sow their crops and when to bring in the livestock to shelter from snow and icy winds. Fishermen trusted their lives to their ability to predict the approach of storms. Weather and its prediction mattered.

This chapter describes the beginning of the process by which weather prediction changed from folklore to science. It also tells the story of the discovery of the atmospheric gases and the answer to a question every child asks: Why is the sky blue?

ARISTOTLE AND THE BEGINNING OF METEOROLOGY

Meteorology is the scientific study of the weather. The scientist who practises meteorology is a meteorologist. The word *meteorology* is derived from two Greek words: *meteoros,* meaning "lofty," and *logos,* meaning "word" or "account." So the Greek word *meteorologia* means "account of lofty [atmospheric] phenomena." Aristotle was the first person to use the word *meteorologia* in a written work that has survived, and the modern science of meteorology can trace its name all the way back to *Meteorologica,* a book he wrote in about 350 B.C.E.

Aristotle (384–322 B.C.E.), a Greek philosopher, was one of the most original thinkers the world has ever seen. Everything interested him, and he wrote an estimated 150 books, of which 30 have survived. Some of these are very short, but others comprise several volumes. Many consist of what appear to be lecture notes that he might have used when discussing matters with his pupils. Aristotle wrote about logic, ethics, politics, aesthetics, biology, physics, astronomy, and many other subjects. He founded the science of zoology, classifying animals into genera and species (although he did not use these terms in the way biologists use them today), and wrote detailed descriptions of many animals.

Aristotle was born in 384 B.C.E. at Stagirus, a Greek colony on the coast of Macedon (modern Macedonia). The map shows the territories of the Mediterranean region as they were during Aristotle's lifetime. Most of the region came to be ruled by the Macedonian

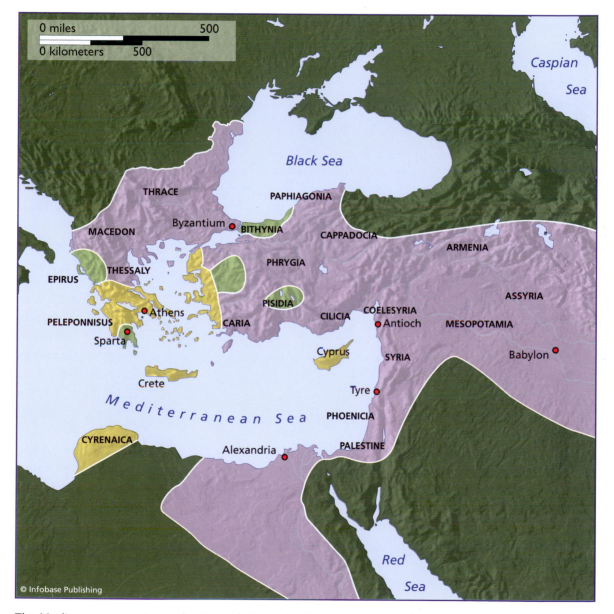

The Mediterranean region in the time of Alexander the Great. At its peak, Alexander's empire covered Egypt and extended eastward as far as modern Rajasthan, India.

king Alexander the Great (356–323 B.C.E.). Alexander's expansion occurred during Aristotle's lifetime.

Aristotle's parents were Greek. His father, Nichomachus, was a physician at the royal court and personal physician to the king, Amyntas III, and medicine was the first subject Aristotle studied. Nichomachus died while Aristotle was still a boy, and a guardian named Proxenus assumed responsibility for his upbringing. In about 367 B.C.E., when Aristotle was 17, Proxenus sent the young man to Athens, to enroll at the Academy, the school led by the philosopher Plato (428 or 427–348 or 347 B.C.E.). Aristotle remained there until Plato's death. By that time King Amyntas had died and been succeeded by his son, Philip II, and Athens and Macedon were at war. Although he was Greek, Aristotle was sympathetic to the Macedonian cause, which would have made him unpopular. Perhaps for that reason, or because he saw no point in remaining once Plato was dead and he had not been chosen to succeed him, Aristotle left Athens. For a time he settled on the coast of Asia Minor (modern Turkey) and then moved to the island of Lesbos in the Aegean Sea, where he lived from 345 B.C.E. until 343 B.C.E., when he returned to Macedon. Philip II appointed Aristotle to supervise the education of his 13-year-old son, Alexander. Later in his life, Aristotle was very wealthy, possibly because Philip paid handsomely to have his son educated by such an impressive tutor. Alexander's formal education was interrupted by military campaigns and in 336 B.C.E., when his father was assassinated, Alexander became king at age 16 and his lessons ended.

Aristotle returned to Athens in about 335 B.C.E., and for 12 years he taught at the Lyceum, a school close to the temple of Apollo Lyceus, from which it derived its name. Alexander, meanwhile, had extended his empire across the known world and had become Alexander the Great. When he died in 323 B.C.E., people with Macedonian connections once more became unpopular in Athens. Aristotle was friendly with a Macedonian general and charged with impiety. Rather than face the possibility of execution, he moved to Chalcis (modern Khalkis) on the island of Euboea, where he was safe. The following year he fell ill and died. He was 62.

In *Meteorologia* Aristotle discusses events that "take place in the region nearest to the motion of the stars," and he includes "all the affections we may call common to air and water, and the kinds and parts of the earth and the affections of its parts" (Book I, Part 1). This leads to an explanation of a variety of phenomena including thunderbolts, winds, earthquakes, and whirlwinds. Aristotle believed that all bodies that move in a circle owe their existence and motion to four principles: fire, air, water, and earth. These form concentric spheres, with earth at the bottom surrounded by water, which is surrounded by air, and finally by fire. Aristotle explains what this implies for the structure of the region between the Earth and stars. Aristotle recognized that the heat of the Sun evaporates water. "The exhalation of water is vapor: air condensing into water is cloud. Mist is what is left over when a cloud condenses into water, and is therefore rather a sign of fine weather than of rain." "So the moisture is always raised by the heat and descends to the earth again when it gets cold. These processes and, in some cases, their varieties are distinguished by special names. When the water falls in small drops it is called a drizzle; when the drops are larger it is rain" (Book I, Part 9). Cooling produces rain, and also snow and hail. Snow and hoarfrost, says Aristotle, are the same thing, and so are rain and dew; "only there is a great deal of the former and little of the latter" (Book I, Part 11).

Aristotle ends Book I by explaining the origin of rivers and devotes most of Book II to explaining the sea. He proposes that salt water is heavy and what he calls "drinkable, sweet water" is light. The light water is drawn upward, to fall as rain, leaving the heavy salt water behind in the lowest places, where it accumulates (Book II, Part 2). The sea is salty, he suggests, because of the "admixture of something earthy with the water." Aristotle also observes that salt water is denser than freshwater: "ships with the same cargo very nearly sink in a river when they are quite fit to navigate in the sea" and "there is a lake in Palestine, such that if you bind a man or beast and throw it in it floats and does not sink . . . this lake is so bitter and salt that no fish live in it and . . . if you soak clothes in it and shake them it cleans them" (Book II, Part 3). Book III explains rainbows, mock suns, haloes, and other optical phenomena. He states that rainbows are caused by the reflection of sunlight from water droplets, its colors being due to the effect of the reflection passing through air. Book IV discusses the four *principles* (often called *elements*) in more detail.

Although Aristotle's work is by far the most influential, it was built on the ideas of earlier Greek philosophers. Anaximander (610–546 B.C.E.) of Miletus also questioned traditional explanations. He asserted that the wind is air that masses together and is set in motion by the Sun, rain comes from vapor rising from things beneath the Sun, and that thunder and lightning have natural explanations. He did not believe that Zeus hurls thunderbolts at the Earth.

Scientists no longer believe in the four principles, or elements. Aristotle was mistaken in this view, but many of his explanations for meteorological phenomena were not far removed from the modern view of them. This is remarkable, because Aristotle possessed no instruments with which to measure atmospheric conditions and he had no way of entering and directly experiencing the atmosphere above ground level. His strength—and his greatest contribution to intellectual development—was his insistence on basing all his explanations on direct observation and the power of his logic. Following philosophers such as Anaximander, Aristotle taught his pupils and followers never to accept an explanation simply because it was the traditional view or because it was what the authorities or important people believed. They must carefully consider explanations advanced by others, but accept them only if they made sense and were in accordance with observation. Wind, rain, snow, storms, floods, droughts, and all the other aspects of the weather are not produced by the whims of gods who are easily offended and as easily appeased.

That is how the science of meteorology began. It was a solid base, not because Aristotle was correct, but because he demonstrated that the weather is natural and results from natural forces that can be understood.

THEOPHRASTUS AND WEATHER SIGNS

Theophrastus (371 or 370–288 or 287 B.C.E.) also studied philosophy under Plato at his Academy in Athens, at the time when Aristotle taught there. Following Plato's death, Theophrastus may have accompanied Aristotle on his travels (see "Aristotle and the Beginning of Meteorology" on pages 2–6). In any event, it is known that he later became one of Aristotle's pupils at the Lyceum. Aristotle liked to walk as he talked, so his lectures and discussions took place as he and his audience strolled along a covered walkway called the *peripatos*, which Aristotle had had built at the Lyceum, giving it the nickname of the Peripatetic school. Theophrastus was Aristotle's favorite pupil.

Indeed, it was Aristotle who gave him the nickname Theophrastus, meaning divine speech; his real name was Tyrtamus. This photograph is probably a realistic likeness.

Tyrtamus or Theophrastus was born at Eresus on the island of Lesbos. When Aristotle retired or fled to Chalcis in 322 or 321 B.C.E., he handed over the library and all the manuscripts at the Lyceum to Theophrastus, making him the head of the Peripatetic school. Theophrastus earned his nickname, for he was an extremely popular teacher, attracting students from far and wide. During his tenure the Lyceum had as many as 2,000 students. He also had the support of the Macedonian kings Philip II (382–336 B.C.E.), Ptolemy (367–283 B.C.E.), and Cassander (ca. 358–297 B.C.E.). So great was the esteem in which Theophrastus was held that when the authorities tried him for the capital offense of impiety the Athenian jury refused to convict him. He died in 288 or 287 B.C.E., having headed the Lyceum for 35 years, and was given a public funeral attended by a large number of Athenians.

Theophrastus (371 or 370–288 or 287 B.C.E.) followed Aristotle as head of the Lyceum in Athens. Although he was best known as a botanist, he was also one of the founders of meteorology. *(Getty Images)*

Like his teacher, Theophrastus had wide interests and wrote on many subjects. He is often described as the founder of botany because of his books *Enquiry into Plants* and *On the Causes of Plants*, but in about 300 B.C.E. he also wrote two books on meteorology: *On the Signs of Rain, Winds, Storms, and Fair Weather* and *On Winds*. He had studied earlier writers on the subject, and he obtained information from farmers and from sailors who plied the Aegean, so his meteorology was based largely on the accounts of others.

Most of the observations Theophrastus collected were reliable, and his explanations for them were usually accurate. He believed that wind is air in motion and he noted, correctly, that in Greece the strongest winds are those blowing from the north and south. He proposed that their strength and warmth varied according to the distance the winds had traveled and the terrain they had covered. The winds also varied with the seasons. He quoted an Athenian saying, also mentioned by Aristotle: "North winds blow in the summer, and in late autumn until the end of the season, while the south winds blow in winter, at the beginning of spring, and at the end of late autumn." The west wind, which blows only in spring and late autumn, is sometimes mild, but at other times can destroy crops. This, he says, is because the air has traveled across the sea.

Theophrastus was familiar with mountain weather. "When the winds blow against the high mountains near Olympus and Ossa and do not surmount them, they lash back in the reverse direction, so that the clouds moving on a lower level move in reverse direction." He also wrote that winds blowing down the mountainsides often produced squalls over the sea. The approach of winds can be predicted by interpreting the signs that gave Theophrastus the title for his work. At sea, the surface waves provide information, as does the behavior of dolphins and other marine animals. Useful sky signs include haloes, mock suns (parhelia), and shooting stars.

Aristotle speculated on the cause of winds. Theophrastus was more cautious, although he believed the Sun, Moon, and stars exerted an influence. He suggested that as it rises the Sun sets the winds in motion, but also stops them. The Moon does the same, but the effect is weaker because the Moon itself is weaker.

Because Theophrastus relied heavily on what he heard from farmers, sailors, and others with a particular interest in the weather, he became aware that climates change over time. He reported that Crete suffered severe winters, with heavy snow, but said that long ago the climate was much milder and the mountain slopes, barren in his day, could be cultivated. Information of this kind can have reached him only from a kind of folk memory of local people; neither he nor they had anything a modern scientist would accept as evidence to support their beliefs. Exceptional winters and summers imprint themselves on the memory while average seasonal weather is forgotten. That is why elderly people often suppose the weather was markedly different when they were young. Although many of his reports and interpretations were sound, Theophrastus was recounting weather lore—hearsay. Weather lore consists of descriptions of natural signs that are believed to predict the weather (see the sidebar), often expressed as short rhymes or popular sayings. Some are reliable, but most are not.

Nevertheless, Theophrastus did much more than repeat the teachings of Aristotle and gather folk beliefs. He built on Aristotle's work, disagreed with his predecessor over certain details, and directed his own followers to base their understanding of the weather on observations and accounts that led to natural explanations. He fully deserves to be regarded as one of the founders of atmospheric science.

WEATHER LORE

People have always tried to predict the weather, usually for very practical reasons. Farmers need to know whether there will be a late or early frost, fishermen whether there will be a storm, and travelers whether the clouds they see mean they should hurry to seek shelter. But until scientists learned how to forecast the weather, predictions had to be based on experience and the signs of approaching weather they could see around them. Sailors knew, for example, that mares' tails—wisps of cirrus cloud, curling at the ends—meant the wind would soon strengthen, and they were usually right. Over the centuries these signs accumulated into a large body of weather lore encapsulated in sayings and rhymes.

Some of the rhymes are well known and often reliable. These include:

> Red sky at night, shepherd's delight.
> Red sky in the morning, shepherd's warning.

This is often true, as are:

> Rain before seven,
> Fine before eleven.

and

> Dew in the night
> Next day will be bright.

Summer mornings often begin with a thin mist—in fact, a shallow layer of *radiation* fog—that evaporates (people say it burns off) as the Sun rises and the air warms. Hence:

> Gray mists at dawn,
> The day will be warm.

Other sayings are based on observations of animals. Northerners say that one swallow doesn't make a summer. This refers to barn swallows, migratory birds that winter in the south and spend summer in the north. They do not all arrive together, so the appearance of a few individuals, probably swept north on a strong wind, does not mean summer has arrived. In Britain people say Ne'er cast a clout till May be out. A clout (cloth) refers to winter underwear, and it is not clear whether May is the month or May blossom, the flowers of hawthorn, a familiar plant of hedgerows and roadsides, which open in early summer. It is also said that cows lie down when rain approaches, scratch their ears when a shower is imminent, and gather on top of a hill when the weather will be fine.

There are also beliefs about control days. The weather on a control day predicts the weather for some time afterward.

> If Candlemas be fair and bright,
> Winter'll have another flight.
> But if Candlemas Day be clouds and rain,
> Winter is gone and will not come again.

Candlemas (February 2) is a religious festival that is traditionally celebrated with lighted candles, and this rhyme belongs to the same tradition as Groundhog Day, which is also February 2. That is the day when, in parts of North America, the groundhog emerges from its burrow where

(continues)

(continued)

it has spent the winter in hibernation and looks for its shadow. If it sees the shadow (showing the day is sunny), the groundhog retreats into its burrow and stays there for a further six weeks. If it cannot see its shadow, it remains in the open.

Many control days are religious festivals, because these are dates people remember. Easter provides several predictions, including:

> Easter in snow, Christmas in mud;
> Christmas in snow, Easter in mud

and

> If it rains on Easter Day,
> There shall be good grass but very bad hay.

Such long-range predictions are unreliable, but who will remember at Christmas what the weather was like last Easter? Others, however, are based on straightforward observation. If Easter Day is rainy, the rain will encourage the grass to grow, but will also make it difficult to dry mown grass to make hay.

People still repeat some of the old sayings, but they cannot compete with the colored maps, symbols, and self-confidence of the television weather forecaster. It will be sad, though, if this ancient weather lore is completely lost.

JAN BAPTISTA VAN HELMONT AND THE DISCOVERY OF GASES

The Greeks believed that nature was regulated by four elements or principles: earth, water, air, and fire. These elements were not material substances. The element earth was not made of soil or rock, and air was not the mixture of gases that we understand it to be. The words had different meanings, reflecting a radically different view of the natural world. That view was strongly influenced by Pythagoras (ca. 569–ca. 475 B.C.E.). Pythagoras is famous today for the theorem bearing his name, but as well as being a mathematician he was a philosopher and religious leader.

A school of philosophy founded by Pythagoras flourished in Greece in about 500 B.C.E. Its members were known as the *Pythagorean Brotherhood,* and they believed that the Earth, planets, Sun, and Moon (as well as the invisible Anti-Earth, hidden on the far side of the Sun) were set on spheres of crystal that rotated around a central fire. Movement of the spheres produced harmonious music—the music of the spheres. The Pythagoreans also believed that everything is formed from whole numbers and geometric shapes. Certain shapes were of particular interest to them because of their mathematical regularity. A plane (two-dimensional) geometric shape such as a triangle, square, or rectangle is called a *polygon*. A three-dimensional solid figure with faces that are plane polygons is a *polyhedron*.

The Pythagoreans knew of four polyhedra, which they associated with the four principles (elements). The bottom diagram in the illustration that follows shows how these were arranged, with the tetrahedron representing fire, the hexahedron representing earth, the icosahedron representing water, and the octahedron representing air. The conditions hot, moist, cold, and dry were determined by the influences of these elements. Clearly, the concept of each of the elements was very unlike the meaning their names have today. This becomes still more evident when the Greeks realized that there are five possible polyhedra, not four. When the fifth polyhedron, the dodecahedron, was discovered, the Pythagoreans found it necessary to add a fifth element for it to represent—a *quinta essentia*, from which we derive the word *quintessence*. They called this the aether. Centuries later the five polyhedra became known as the Platonic solids, in honor of the philosopher and teacher Plato (428–348 or 347 B.C.E.) who mentioned them in his writings. The five polyhedra are shown in the top drawing in the illustration.

For the ancient Greeks, air existed as a principle or basic element. Approximately 2,000 years passed before scholars learned that air is a distinct, material substance with its own properties and made from gases. The first step in that discovery was made in the early 17th century by the Flemish physician and chemist Jan Baptista van Helmont (1577–1644).

Van Helmont grew up in the tradition of the four principles, but he rejected Aristotle's ideas about them, asserting instead that fire is not an element and neither is earth, because earth can be reduced to water. That left only water and air, and he maintained that air is merely a physical matrix or structure that contains various substances but does not react with them. Only water undergoes chemical change. He found biblical support for this view in Genesis, and he proposed a series of processes by which water was transformed into every other material substance.

In order to test his belief that everything is made from water, van Helmont carefully weighed a young willow tree before planting it in a pot containing a quantity of soil he had also weighed. He kept the tree in its pot for five years, nourishing it only with water. At the end

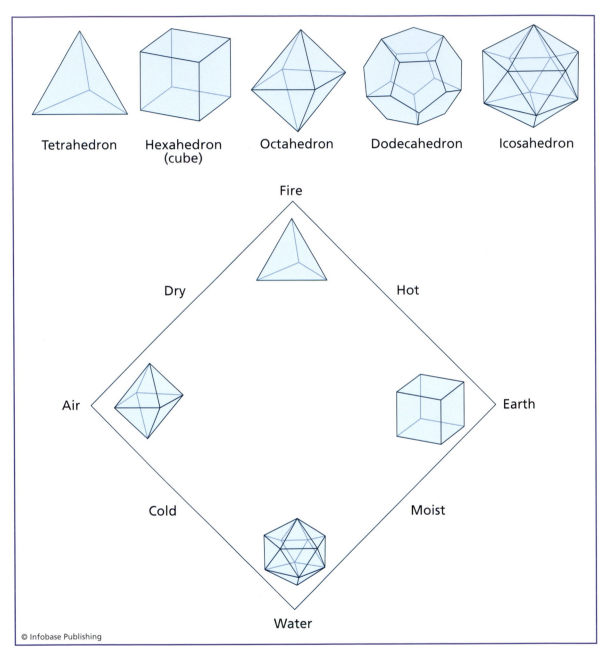

Five three-dimensional shapes can be constructed with faces that are regular geometric forms. These are known as Platonic solids, in honor of Plato. (*Top*) These shapes, called polyhedra, are named by the number of their faces: tetrahedron (4); hexahedron (6); octahedron (8); dodecahedron (12); and icosahedron (20). (*Bottom*) The ancient Greeks associated four of these solids with the four principles (elements).

of five years he weighed the tree and soil again. He found that the tree had gained 164 pounds (74.5 kg), but the soil had lost only 2 ounces (57 g). He took this as evidence that the tree had transformed water into its own substance. (He was mistaken, of course, but photosynthesis had not yet been discovered.)

By the 17th century students of the natural world were performing controlled experiments, and educated people everywhere followed their work and discoveries with keen interest. Van Helmont was meticulous. He measured everything carefully and observed the results accurately. Some of his experiments released vapors, which he recognized as substances, each with its own properties. Once released, these substances expanded rapidly to fill any container they entered. He took this to mean that they existed in a formless state he described as chaos, using the Greek word *khaos*. He wrote *khaos* the way he pronounced it, as *gas*.

He had coined the word *gas*, but van Helmont did not use it in quite the modern sense. He thought that every gas was made from the same material, but modified by the processes it had undergone, and that every natural substance contains a gas, which can be released if the substance is heated. To demonstrate this, van Helmont burned 62 pounds (28 kg) of charcoal and was left with 1 pound (0.5 kg) of ash. He assumed the remaining charcoal had been released as a gas, which he called "gas sylvestre," from *sylva*, the Latin word for wood. He found the same gas given off when wine and beer are fermented. (Its modern name is carbon dioxide.)

Jan Baptista van Helmont was born into an aristocratic family in Brussels on January 12, 1577. He studied Latin and Greek at the University of Louvain, but refused to take a degree because he believed such honors were mere vanity. He attended courses held by Jesuit teachers, studied mystical Christian writers, and finally turned to medicine, qualifying as a physician in 1599. He spent the next few years traveling in Switzerland, Italy, France, and England, before returning to Flanders (now Belgium) and settling first in Antwerp. In 1609 he married a wealthy wife, Margaret van Ranst. After his marriage he moved to an estate in Vilvorde, near Brussels, where he lived for the rest of his life. He died at Vilvorde on December 30, 1644.

In 1621 van Helmont published a paper, *De magnetica vulnerum curatione*, describing the use of magnetism to cure wounds. This defended the views of a Calvinist professor, Rudolf Goclenius, against those of a Jesuit, Johannes Roberti, and it was thought to challenge certain miracles. The paper attracted the attention of the Inquisition, which brought charges against him. He was under house arrest from 1633 until 1636 and not allowed to publish anything, and he was not acquitted until 1646—two years after his death. The experience made him reluctant to publish his other papers. These were assembled later by his son Franz Mercurius van Helmont (1614–99) and published posthumously in 1648 as *Ortus medicinae* (Origins of medicine).

Van Helmont bridged two intellectual worlds. On one hand he inherited from his medieval predecessors the Greek concept of the elements. He was an alchemist, who claimed to have seen the philosopher's stone, the substance alchemists believed would transmute base metals such as lead into gold. At the same time he challenged old ideas and tested his own theories by experiments that he conducted rigorously.

KARL SCHEELE, JOSEPH PRIESTLEY, AND DEPHLOGISTICATED AIR

On December 9, 1742, Karl Wilhelm Scheele (1742–86) was born at Stralsund, Pomerania. At that time the Duchy of Pomerania, bordering the Baltic Sea, was part of Sweden; today it lies partly in Germany and partly in Poland. Despite being Swedish by nationality, Scheele spoke German. The 18th century was an age of rapid scientific advance, and Scheele was one of the most talented chemists of his generation.

His father was a carpenter and Karl was the seventh of 11 children. The family was poor, and there was no money to pay for his education, but when he was 14 Karl was apprenticed to Martin Anders Bauch, owner of an apothecary (druggist) company in the city of Göteborg. Karl had learned to read and write, and he was a keen observer and very eager to learn chemistry. He read books on the subject and performed experiments, for which he had a natural talent. His apprenticeship lasted eight years, and in 1765 he became a clerk to an apothecary in Malmö, where he remained until 1768 when he obtained a more responsible position as an apothecary in Stockholm. He worked in Uppsala from 1770 until 1775, then moved to Köping, in Västerland, where in 1776 he was able to open his own establishment and where he remained for the rest of his life. Later he married the widow of the previous owner, but by that time he was gravely ill, his health weakened by constant overwork and his unfortunate habit of tasting the products of his chemical experiments. Karl Scheele died two days after his wedding, on May 21, 1786.

Scheele's talent was widely recognized. He was elected to the Swedish Royal Academy of Sciences in 1775, but refused lucrative offers of employment in England and an invitation from Frederick II of Prussia to become his court apothecary in Berlin. He preferred life in a small town, although he remained poor all his life and his house was bitterly cold and drafty during the long Swedish winter.

He wrote only one book, *Chemische Abhandlung von der Luft und dem Feuer* (Chemical treatise on air and fire), published in 1777. In it he described two of his discoveries, of oxygen, which he called "fire air," and nitrogen. He also discovered barium (1774), manganese (1774), chlorine (1774), molybdenum (1778), and tungsten (1781), as well as a long list of compounds, including hydrogen fluoride, silicon fluoride, tartaric acid, glycerol, citric acid, lactic acid, uric acid, benzoic acid, hydrogen sulfide, and many more. Copper arsenite, another of his discoveries, is still known as Scheele's green. Hydrogen cyanide, arsenic, and mercury compounds were among the many substances he tasted, and the symptoms of his final illness resembled those of mercury poisoning.

Scheele discovered chlorine by heating the manganese mineral pyrolusite (MnO_2) with hydrochloric acid (HCl). The mixture gave off a thick, yellowish gas that sank downward. It would not dissolve in water and it bleached the color from litmus paper and from some flowers. He called the gas dephlogisticated marine acid.

The chemistry Karl Scheele studied explained that combustible substances contained *phlogiston,* which combustion released (see sidebar). That is how he came to describe chlorine (the name given to it by Sir Humphrey Davy, 1778–1829) as dephlogisticated.

Scheele discovered fire air some time prior to 1773, and he used several methods to produce it. By the time he published the fact in 1777, however, credit for the discovery had already been given to the English chemist Joseph Priestley (1733–1804). Both Karl Scheele and Joseph Priestley used experimentation to advance chemical knowledge, following in the tradition established by van Helmont and others, and the understanding that arose from their work led directly to the total rejection of the phlogiston theory. It is ironic, therefore, that

Priestley clung tenaciously to that theory to the end of his life and defended it vigorously.

Joseph Priestley was much more than a talented chemist. He was a minister of religion who helped found Unitarianism, a theologian, a linguist, an educator who published an important and progressive

PHLOGISTON

In 1667 the German chemist Johann Joachim Becher (1635–82) published a book called *Physica Subterranea* (Physics below ground), in which he revised the traditional view of the classical elements. Becher replaced the elements fire and earth with three alternative forms of earth to which he gave Latin names: *terra lapidea,* or "stony earth," which was the quality allowing earth to fuse into a solid mass; *terra fluida,* or "flowing earth," governing the ease with which earth will flow; and *terra pinguis,* or "fatty earth," which is concerned with combustion. Becher argued that when any combustible substance is burned *terra pinguis* is released.

One of Becher's students was another German chemist, Georg Ernst Stahl (1660–1734). Stahl expanded Becher's ideas and renamed *terra pinguis,* calling it *phlogiston,* from the Greek word *phlogos,* meaning "flame." Phlogiston was colorless, odorless, tasteless, and could not be sensed by touch, but it possessed mass. Every combustible substance contained it and released it when it was burned along with caloric (heat). The residue, after burning, was called calx (plural calces). Calx was the true form of the substance.

Prior to burning, a substance was said to be phlogisticated, and after burning the calx was dephlogisticated. Calx weighed less than the original phlogisticated substance because

it had lost phlogiston. Different substances left behind different amounts of calx depending on the amount of phlogiston they contained. Charcoal and sulfur leave very little calx because they are almost pure phlogiston.

Metals also contain phlogiston and can release it. The process is called calcination. When this happens the resulting calx (rust in the case of iron) weighs less than the original phlogisticated metal. However, some metals can be restored from their calces by heating them with burning charcoal. Phlogiston leaving the charcoal enters the metallic calx, thereby phlogisticating it.

Burning released phlogiston into the air, and when substances were burned in an enclosed space the air became increasingly phlogisticated. A point could be reached where the phlogisticated air was incapable of supporting further combustion. It was saturated with phlogiston. What is more, animals could not survive in fully phlogisticated air, because respiration became impossible. Respiration, therefore, removed phlogiston from the body.

The phlogiston theory was highly successful because for more than a century it provided a plausible explanation for natural phenomena and experimental results. Chemists believed it, and it was not until late in the 18th century that it was finally disproved and quickly abandoned.

textbook on English grammar, and a political theorist who supported the ideals of the American and French Revolutions and the rights of Dissenters—those who refused to accept the doctrines of the Church of England.

Joseph Priestley was born on March 13, 1733, in the small town of Birstall, West Yorkshire, about six miles (10 km) from Leeds, the oldest of the six children of Jonas Priestley, a finisher of cloth, and his wife, Mary Swift. His mother died when Joseph was six, and when his father remarried in 1741 the boy went to live with his wealthy uncle and aunt, John and Sarah Keighley. He attended local schools where he learned Latin, Greek, and Hebrew. He also studied French, German, Italian, Chaldean, Syrian, and Arabic. The family was Calvinist and therefore religious Dissenters, and his education continued at a dissenting academy in Daventry, Warwickshire. He matriculated in 1752, and in 1758 he became a clergyman, with parishes in Needham Market, Suffolk, and later at Nantwich, Cheshire. In 1761 he was appointed to teach modern languages and rhetoric at the dissenting Warrington Academy in Cheshire. Until the laws were repealed in 1829, Roman Catholics and Dissenters, also known as Nonconformists although the terms are not strictly synonymous, were discriminated against in England. They were not permitted to hold public office, stand for election to Parliament, serve in the army or navy, or attend the universities of Oxford or Cambridge.

On June 23, 1762, Joseph married Mary Wilkinson. While at Warrington, Priestley conducted scientific experiments, mainly on electricity, and lectured on anatomy. It was at this time that he met Benjamin Franklin (1706–90), who encouraged his growing interest in science. In 1767 Joseph, Mary, and their daughter, Sarah, moved to Leeds where Joseph became minister of Mill Hill Chapel. They remained there until 1773, and two sons, Joseph and William, were born during their time there.

While he was at Mill Hill, Priestley sent five scientific papers to the Royal Society, describing his experiments with electricity and optics. The Royal Society awarded him their Copley Medal in 1773. Priestley also invented soda water, publishing a description of the method he used for the benefit of the crew sailing on James Cook's (1728–79) second voyage to the South Seas—he mistakenly believed it would cure scurvy. Priestley made no money from soda water, but a

German silversmith and watchmaker called Johann Jacob Schweppe (1740–1821) patented the process in 1783.

In 1772 William Petty, the second earl of Shelburne (1737–1805), offered Priestley a position as librarian and tutor to his children, which would offer the family greater financial security, and the following year they moved to the Shelburne estate at Calne, Wiltshire, where their third son, Henry, was born in 1777. It was at Calne that Priestley did his most important scientific work, concerned mainly with gases, which Priestley always called airs. He discovered nitrous air (nitric oxide, NO), alkaline air (ammonia, NH_3), acid air (hydrochloric acid, HCl), and dephlogisticated (also called diminished) nitrous air (nitrous oxide, N_2O). He later discovered vitriolic acid air (sulfur dioxide, SO_2), and also isolated carbon monoxide (CO), but failed to recognize it as a distinct air. The discovery for which he is best known was of dephlogisticated air, which would later be renamed oxygen. Priestley found that mice survived when placed in a container of dephlogisticated air, so he tried the gas himself, finding it an improvement on ordinary air.

The family moved to Birmingham in 1780, following a disagreement with Lord Shelburne, and remained there until 1791. Priestley became a minister again, and he also joined the Lunar Society, a group of scientists, engineers, inventors, and manufacturers who met once a month when the Moon was full, to minimize the risk of being attacked on the unlit streets. Many of the papers he published at this time were devoted to defending the phlogiston theory.

For many years Priestley had devoted considerable effort to theological and political campaigning. This made him a controversial figure. He claimed in his pamphlets that the teachings of the early Christian church had been corrupted, and he argued strongly for freedom to express dissenting opinions. His language grew increasingly extreme until, at the height of the French Revolution, he appeared to be calling for revolution in England. Their support for the American and French Revolutions had made the Dissenters increasingly unpopular, and in 1791 riots broke out in Birmingham. Joseph and Mary Priestley fled from their home, which was attacked and burned to the ground, destroying all their possessions and Joseph's laboratory. After hiding with friends for several days they escaped to London. They lived at Clapton until 1794, Joseph lecturing on science and history at a dissenting academy, but life became more and more difficult. Cartoons were published attacking him, and an effigy of him was burned. He was forced to resign from the Royal Society, was attacked in speeches in Parliament, denounced by preachers, and his sons were unable to find work. The sons decided to emigrate to America, and, although Joseph had been made an honorary citizen of France, he and Mary decided to accompany them, escaping shortly before the government began arresting those who spoke out against its policies. The Priestleys sailed for America on April 17, 1794, arriving to a warm welcome in New York. They then moved to Philadelphia, where Joseph was offered, but declined, the professorship of chemistry at the University of Pennsylvania. Instead, the family moved to the town of Northumberland, where their son Joseph and others were establishing a colony for English Dissenters.

Henry Priestley died in 1795, and Mary died the following year. Joseph's health deteriorated, and by 1801 he was unable to work. He died at Northumberland on February 6, 1804. He had been elected to the membership of every leading scientific society in the world. Georges Cuvier (1769–1832), the most famous naturalist in Europe, wrote his eulogy, and the German philosopher Immanuel Kant (1724–1804) praised him in his most famous work, *Critique of Pure Reason,* published in 1781. Since 1952, Dickinson College in Carlisle, Pennsylvania, has presented an annual Priestley Award to a scientist whose discoveries contribute to the welfare of mankind.

ANTOINE-LAURENT LAVOISIER AND OXYGEN

Joseph Priestley discovered dephlogisticated air by heating mercuric oxide. This restored the metal, in Priestley's view, by returning to it the phlogiston it had lost when it was calcined (roasted in air) to change the metal to its calx. The air surrounding the heated mercuric oxide was breathable because phlogiston had been removed from it. Hence, the air was dephlogisticated.

Priestley visited Paris in 1774, where he met the French chemist Antoine-Laurent Lavoisier (1743–94). The two became friends, maintained their friendship through correspondence, and Lavoisier learned much from Priestley, for whom he had the deepest respect. That affection made him very reluctant to disagree with Priestley over a matter that was central to Priestley's scientific outlook, but, in the end, disagree he did. Lavoisier found the phlogiston theory was unsustainable and proved that it was wrong.

According to the phlogiston theory, air containing phlogiston was unbreathable and candles would not burn in it, but animals could breathe dephlogisticated air and it supported combustion. Lavoisier suspected a close link between combustion and respiration. He noted that when a metallic calx was heated in the presence of charcoal the gas produced is what was then known as fixed air, but, because the charcoal was destroyed completely, Lavoisier concluded that fixed air was formed by the combination of two gases, one given off by the heated calx and the other by the charcoal. (Fixed air is now called carbon dioxide.) The illustration shows a piece of apparatus Lavoisier actually used. It was made from one of Madame Lavoisier's drawings of her husband's laboratory and equipment.

Lavoisier also heated samples of iron calx (rust) in a bell jar without charcoal, using a burning glass as a source of heat. As the illustration on page 22 shows, the burning glasses chemists used in the 18th century were huge devices that produced high temperatures.

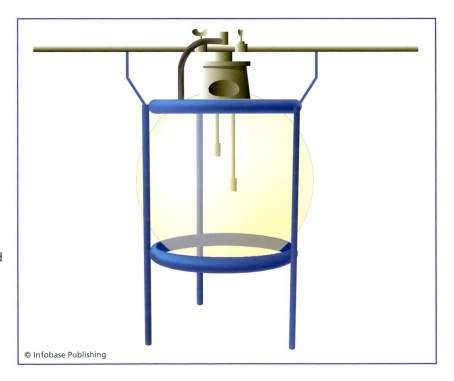

The apparatus Lavoisier used for an experiment on the combustion of hydrogen, reconstructed in 1783 by Jean-Baptiste Meusnier (1754–93) from a sketch left by Lavoisier

© Infobase Publishing

This experiment released large amounts of a gas Lavoisier called elastic air, but he said it mixed rapidly with common air, and he found it impossible to tell whether the properties of the gas in the bell jar belonged to the elastic air emitted by the calx or to the common air or to a mixture of the two. He repeated the experiment with red mercuric oxide, heating the calx in a retort (heating vessel) with a measured amount of charcoal, and found a mouse could not breathe the fixed air and it extinguished a candle. Then he heated another sample of the calx, this time without charcoal. This time charcoal burned brightly and vigorously in the elastic air and a mouse was able to breathe. Lavoisier concluded that the principle that combines with metals and increases their weight when they are calcined is some ingredient present in common air that supports respiration and combustion, and that fixed air is a combination of this portion of common air with some ingredient of charcoal.

Lavoisier's experiments also revealed that phlogisticated air could be dephlogisticated if the retort contained water and was shaken vigorously. Water removes fixed air (because carbon dioxide is soluble in water). Priestley, along with most chemists, believed that when a candle (or anything else) burns, combustion reduces the amount of air. When Lavoisier shook a flask of water and fixed air the volume of air decreased by about one-third. He then went on to discover that a candle would burn in the remaining two-thirds of the air. Clearly some of the fixed air was combining with water. So he repeated the experiment using mercury instead of water, so the "fixed air" could not disappear by combining with water. This time he found the volume of air in the flask remained unchanged. He concluded that combustion changes common air into fixed air.

The experiments eventually began to show that combustion and respiration were similar chemical processes. Lavoisier measured the amount of heat generated by respiration, using a guinea pig. This may have been the first time a guinea pig was used in an experiment, and it was Lavoisier's use of one that forever linked guinea pigs with experimentation. Lavoisier found that respiration generates the same amount of heat as combustion using a similar amount of oxygen, but that the reaction proceeds more slowly. Essentially, respiration is a slow form of the combustion of carbon contained in organic material. Both types of combustion could be explained without supposing the existence of phlogiston. He advanced very cautiously, repeating

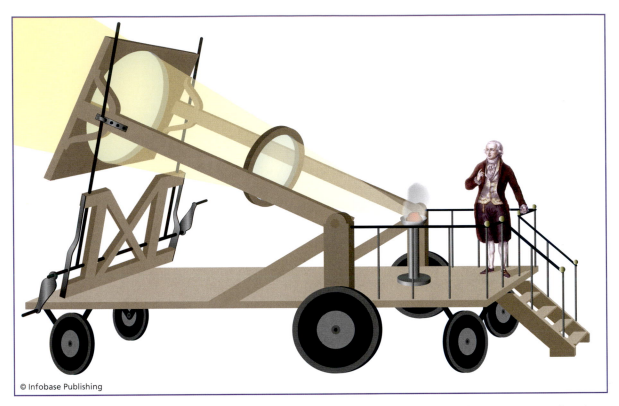

© Infobase Publishing

In order to heat substances without burning charcoal, which would contaminate the gases emitted, 18th-century chemists used burning glasses—lenses that focused sunlight. These were not hand-held toys, however, but very large devices that generated very high temperatures. The picture shows Lavoisier performing one of his experiments.

experiments many times, but eventually he felt ready to propose an alternative to the phlogiston theory. He finally presented his findings in 1777. They demolished the phlogiston theory.

Lavoisier renamed dephlogisticated air as pure air. He then named the ingredient of pure air that made it capable of supporting respiration and combustion, using the Greek word *oxu*, meaning "acid," because he believed (incorrectly) that all acids contain it, and *genes*, meaning "born." The word he coined was *oxygène*. Lavoisier gave oxygen its name, although he did not discover it, but his achievement is more significant than simply coining a name for a gas. The phlogiston theory supposed that air is a single substance that can change its qualities through gaining or losing phlogiston. Lavoisier showed that air is not a single gas, but a mixture of gases.

His work on combustion and respiration formed only a part of Lavoisier's contribution to science. He was the first person to formulate clearly the *law of conservation of mass*. This law states that the mass contained within a closed system will remain constant regardless of the processes occurring inside the system. He introduced the metric system of weights and measures and helped reform the naming of chemical elements and compounds. Most of all, his careful measurement and recording of all the materials and processes in his experiments helped transform chemistry into the scientific discipline it is today. He is often called the father of modern chemistry.

Antoine-Laurent Lavoisier was born in Paris on August 26, 1743. His mother died when he was five, and Antoine inherited a large fortune from her estate. From 1754 to 1761 he studied chemistry, botany, and mathematics at the Collège Mazarin, one of the colleges of the University of Paris, and in 1761 he began studying law, also at the University of Paris, graduating in law in 1763. His interest in science had not decreased, and he continued to attend science lectures during the time he was studying law. He took part in a geological survey of Alsace-Lorraine in 1767 and in a survey of the whole of France in 1769. In 1768 he was elected to the French Academy of Sciences. His father bought him an aristocratic title in 1782.

In 1771 Lavoisier married Marie-Anne Pierrette Paulze, who became a valuable colleague. She translated books and papers from English, including those written by Joseph Priestley and an "Essay on Phlogiston" by Richard Kirwan. Not only did she translate this paper, she added notes of her own pointing out mistakes in Kirwan's chemistry. It was this that convinced Lavoisier that the phlogiston theory was incorrect. Marie-Anne had trained as an artist and drew sketches and prepared engravings of the laboratory and its equipment, and she edited and published Lavoisier's notes and memoirs, including his influential *Elementary Treatise on Chemistry* (1789).

Lavoisier's father-in-law was a member of the Ferme Générale, and in 1768 Lavoisier became an associate and later a full member, one of the 28 official tax collectors. The Ferme Générale was a private organization whose members were required to collect taxes and submit them to the government, each *fermier* being responsible for a particular region of the country. The government stipulated the amount it should receive, but not the amount the *fermiers* could collect, and obviously they had to pay their own salaries and cover their expenses.

Lavoisier in 1790, when he was 47 years of age *(Hulton Archive/Getty Images)*

It was a system wide open to abuse, it was abused, and it was highly unpopular. During the Terror following the French Revolution that began in 1789, all the members of the Ferme Générale came under suspicion, and, in addition, Lavoisier was denounced as a traitor. Some years earlier he had refused to support the attempt by Jean-Paul Marat (1743–93), one of the revolutionary leaders, to become a member of the Academy of Science. Lavoisier had recommended the building of a wall around Paris to control smuggling, and Marat accused him of imprisoning Paris and said that the wall prevented the circulation of air. Furthermore, Lavoisier was known to be corresponding with people—in fact fellow scientists—resident in countries hostile to the Revolution. On May 8, 1794, all the former tax collectors, including Lavoisier, were sent for trial by the revolutionary tribunal. They were all condemned to death, and the same afternoon Lavoisier was guillotined in the Place de la Révolution (now the Place de la Concorde). The illustration shows Lavoisier toward the end of his life.

The following day the Italian-born mathematician and astronomer Joseph-Louis Lagrange (1736–1813) said, *"Cela leur a pris seulement un instant pour lui couper la tête, mais la France pourrait ne pas en produire un autre pareil en un siècle."* (It took them only an instant to cut off his head, but France may not produce another like it in a century.)

As Antoine has come to be called the father of modern chemistry, Marie-Anne, always called Madame Lavoisier, is known as the mother of modern chemistry. After her husband's death, the authorities returned to her all the papers and other items they had seized, with an admission that Lavoisier had been wrongly convicted. Madame Lavoisier collected almost all of his notebooks and apparatus. Most of this material is now held at Cornell University.

DANIEL RUTHERFORD AND NITROGEN

Air is a mixture of gases. The table on page 27 lists its constituents. As the table shows, 78 percent of air consists of nitrogen, with oxygen accounting for almost 21 percent. Joseph Priestley and Antoine

Lavoisier identified oxygen, and in 1772 a young Scottish chemist discovered nitrogen, although he did not name it.

Daniel Rutherford (1749–1819) was the son of John Rutherford (1695–1779), a professor of medicine at the University of Edinburgh, and he became the uncle of the novelist and poet Sir Walter Scott (1771–1832). Daniel was born in Edinburgh on November 3, 1749. He studied medicine at Edinburgh University and qualified as a physician, but he was more interested in chemistry and in plant science. He was appointed Regius Professor of Botany at Edinburgh University in 1786. Regius professorships are created by the British sovereign, and they exist at the universities of Oxford, Cambridge, Dublin, Glasgow, Edinburgh, and Aberdeen. Appointments to all of these with the exception of Dublin have to be approved by the sovereign. In 1786 Professor John Hope (1725–86), keeper of the Royal Botanic Garden, Edinburgh, died, and Rutherford was appointed to the post. Daniel Rutherford held both of these posts until his death in Edinburgh on November 15, 1819.

Rutherford's interest in chemistry was stimulated by one of his teachers, Joseph Black (1728–99; see "Joseph Black, Jean-André Deluc, and Latent Heat" on pages 109–113). Black was studying the properties of fixed air (carbon dioxide). When he sealed a burning candle inside a bell jar, after a time the flame would be extinguished, owing to the accumulation of carbon dioxide. If, without permitting any outside air to enter the bell jar, he dissolved the carbon dioxide in a liquid to remove it, the candle still would not burn. Clearly some other ingredient was preventing combustion. He turned the problem over to his student.

Daniel Rutherford kept a mouse in a sealed container until it died. He then burned a candle, followed by a piece of phosphorus, in the container until those, too, were extinguished. He assumed that the air inside the container would contain an amount of carbon, due to respiration by the mouse, and he passed the air through a strongly alkaline solution to remove the carbon dioxide. Flames would not burn in the air remaining in the container and a mouse placed in the container quickly died.

In 1772 Rutherford reported his results and the conclusions he drew from them. Both he and Black believed in the phlogiston theory—it would be five more years before Lavoisier published his paper proving the theory wrong—and Rutherford explained his work in the language of that theory. Respiration by the mouse and the combustion of the candle and phosphorus released fixed air and phlogiston. Bubbling the air through the alkaline solution removed the fixed air. Rutherford concluded that the remaining air was saturated with phlogiston. He called it fixed air or noxious air. Joseph Priestley, who was also studying the same gas at about that time, called it burnt air or phlogisticated air. Karl Scheele (see "Karl Scheele, Joseph Priestley, and Dephlogisticated Air" on pages 14–19) and Henry Cavendish (1731–1810; see "Henry Cavendish and the Constant Composition of Air" on pages 34–36) were also studying the gas.

In 1789 Antoine Lavoisier proposed calling the gas *azote,* from the Greek *azutos* meaning lifeless, because it will not support life. *Azote* is still its name in French and several other languages. In German it is *Stickstoff,* from *ersticken,* to suffocate, and *Stoff,* material.

It is also an ingredient of niter, also known as saltpeter, which is potassium nitrate (KNO_3), an ingredient of gunpowder. In his *Éléments de chimie* (Elements of chemistry), published in 1790, the French chemist Jean-Antoine Chaptal (1756–1832) invented a new name. He called it *nitrogène,* from the Greek *nitron,* niter, and *geinomai,* to engender. *Nitrogen* is its name in English.

JOHN DALTON AND WATER VAPOR

Water is made from molecules, each molecule comprising two atoms of hydrogen bonded to one atom of oxygen. It is written as H_2O, or sometimes as HOH, which gives a slightly better idea of the way the atoms in the molecule are arranged. The idea that every substance is made from minute particles called atoms was first proposed in about 420 B.C.E. by the Greek philosopher Democritus (ca. 460–ca. 370 B.C.E.), who stated that: "The only existing things are the atoms and empty space; all else is mere opinion." Democritus was speculating, however, and although the idea of atoms was discussed many times over the centuries, the first person to obtain experimental evidence for it and to develop an atomic theory of matter was the English chemist, physicist, and meteorologist John Dalton (1766–1844), and it was his interest in atmospheric gases that led him to it.

On October 21, 1803, Dalton published a paper: "Absorption of Gases by Water and Other Liquids," in which he described what is

COMPOSITION OF THE PRESENT ATMOSPHERE

GAS	CHEMICAL FORMULA	ABUNDANCE
Major constituents		
Nitrogen	N_2	78.08%
Oxygen	O_2	20.95%
Argon	Ar	0.93%
Water vapor	H_2O	Variable
Minor constituents		
Carbon dioxide	CO_2	367 p.p.m.v.
Neon	Ne	18 p.p.m.v.
Helium	He	5 p.p.m.v.
Methane	CH_4	2 p.p.m.v.
Krypton	Kr	1 p.p.m.v.
Hydrogen	H_2	0.5 p.p.m.v.
Nitrous oxide	N_2O	0.3 p.pm.v.
Carbon monoxide	CO	0.05–0.2 p.p.m.v.
Xenon	Xe	0.08 p.p.m.v.
Ozone	O_3	Variable
Trace constituents		
Ammonia	NH_3	4 p.p.b.v.
Nitrogen dioxide	NO_2	1 p.p.b.v.
Sulfur dioxide	SO_2	1p.p.b.v.
Hydrogen sulfide	H_2S	0.05 p.p.b.v.

Note: p.p.m.v. = parts per million by volume; p.p.b.v. = parts per billion by volume

now known as *Dalton's law of partial pressures.* Air, or any other mixture of gases, exerts a pressure. Dalton's law states that the total pressure the mixture exerts is equal to the sum of the pressures each ingredient of the mixture exerts individually. This is known as the *partial pressure* of each ingredient, and it is proportional to the amount of that ingredient the mixture contains. For example, air is approximately 78 percent nitrogen and 21 percent oxygen, and the average sea-level atmospheric pressure is 100 kilopascals (kPa) or 1 bar. Consequently, the partial pressure of nitrogen is 78 kPa (780 millibars) and that of oxygen is 21 kPa (210 millibars). In mathematical terms Dalton's law is expressed as follows:

$$P_{total} = \sum_{i=1}^{n} p_i, \text{ or } P_{total} = p_1 + p_2 + \ldots + p_n$$

where p is the partial pressure of each gas in the mixture.

Dalton had been studying *evaporation, condensation* (see sidebar pages) and the pressure exerted at different temperatures by water vapor and other gases. He found that any gas—he called it an elastic fluid—can be reduced to a liquid by lowering the temperature or increasing the pressure sufficiently. He also noted that the pressure exerted by every gas or mixture of gases changes at the same rate for a similar change in temperature. In fact, this relationship between temperature and pressure is implied by Mariotte's version of *Boyle's law* and by *Charles's law* (see "Robert Boyle, Edmé Mariotte, and Their Law" on pages 101–104 and "Jacques Charles and His Law" on pages 104–106). Dalton's studies of water, however, led him to a puzzle. Water consists of hydrogen and oxygen, but why, Dalton wondered, is the mass of water not shared equally by the hydrogen and oxygen? In his 1803 paper on the absorption of gases he wrote: "Why does water not admit its bulk of every kind of gas alike? This question I have duly considered, and though I am not able to satisfy myself completely I am nearly persuaded that the circumstance depends on the weight and number of the ultimate particles of the several gases."

This line of reasoning led Dalton to develop an atomic theory of matter based on the following five postulates:

1. Matter consists of atoms, which are particles that cannot be reduced to smaller particles.
2. All the atoms of a given element are identical in every respect.
3. The atoms of each chemical element are different from those of every other element, and they are especially different in weight.
4. Atoms cannot be destroyed; chemical reactions merely involve the rearrangement of atoms.

EVAPORATION AND CONDENSATION

Evaporation is the change of phase from liquid to gas; it is the change that occurs when liquid water changes to water vapor, which is an invisible gas (and not to be confused with steam, which consists of liquid droplets that have condensed from water vapor). Condensation is the reverse phase change, from a gas to a liquid; in the case of water from water vapor to liquid water.

A water molecule (H_2O) consists of two hydrogen atoms bonded to one oxygen atom. The angle between lines drawn from the center of the oxygen atom to the centers of the hydrogen atoms is 104.5°. Because of the way the three atoms are arranged and share electrons, the molecule possesses a small positive electromagnetic charge at the hydrogen end (written as H^+) and a small negative charge at the two oxygen ends (O^-). Although the molecule is electromagnetically neutral overall, it carries a positive charge at one end and a negative charge at the other. Molecules carrying charge in this way are said to be polar.

A *hydrogen bond* forms between the hydrogen (H^+) of one water molecule and the oxygen (O^-) of another. In the liquid phase groups of molecules linked by hydrogen bonds constantly form, break apart, and re-form. The molecule groups are drawn toward each other by an attractive force acting equally from all directions on each group, and the groups are able to move freely past each other. At the surface, however, the attractive force acts to the sides and downward, but there is no force acting in an upward direction. Consequently, molecule groups are held at the surface and cannot escape from the main mass of water.

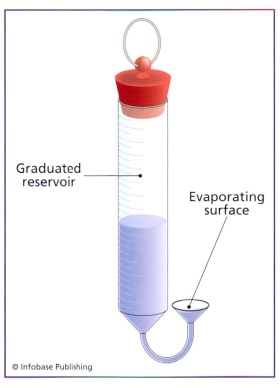

Graduated reservoir

Evaporating surface

© Infobase Publishing

An evaporimeter measures the rate of evaporation. Water is held in a graduated reservoir that connects at its base to a U-tube, widening to a funnel at the top. A layer of filter paper covers the surface of the funnel. As water evaporates from the filter paper in the funnel, the level in the reservoir falls. This is a Piché evaporimeter.

All molecules vibrate, and, if energy is applied (as heat), water molecules vibrate more rapidly as they absorb it. With sufficient energy, water molecules at the liquid surface vibrate so vigorously *(continues)*

(continued)

they break away from their groups and escape into the air. This is evaporation.

Immediately above the liquid surface there is a thin layer of air in which water molecules are constantly entering and returning to the liquid. If the molecules have enough energy to escape beyond this layer the amount of liquid water will decrease; the water will evaporate.

If the free water molecules moving through the air lose energy because the temperature falls they will form new hydrogen bonds with other molecules that approach them. In this way the molecules will join as groups, and groups of molecules will accumulate to form liquid droplets. This is condensation.

The rate at which evaporation or condensation occurs depends on the temperature, because that is a measure of the amount of heat energy available to the molecules.

An *evaporimeter* is used to measure the rate of evaporation. There are several designs. The illustration shows a simple version. It has a graduated reservoir to hold water. The reservoir is sealed tightly at the top. At the bottom it is connected to a U-tube ending in a funnel lined with filter paper. As water evaporates from the filter paper more water is drawn through the U-tube, and the level in the reservoir falls. The rate of evaporation is the amount by which the water in the reservoir diminishes during a measured lapse of time.

5. The formation of chemical compounds occurs through the formation of compound atoms (i.e., molecules), containing a small but definite number of atoms from each element.

Dalton then added a rule of greatest simplicity. This stated that when two elements formed only one compound, a compound atom of that compound invariably contained one atom from each element. Better analytical techniques later showed this to be incorrect, but it allowed chemists to compare the relative weights of different atoms, and Dalton extended it to a law of multiple proportions, stating that when elements combine they do so in proportions of simple whole numbers. In compounds of elements A and B, for example, AB_2 (= A + 2B) contains exactly twice as much B as does AB (= A + B).

From there Dalton devised a system of symbols to represent atoms of different elements, which he published in 1808 in a book *New System of Chemical Philosophy.* The following illustration shows his symbols for elements and for molecules of certain compounds. Dalton classified elements as simple because they consisted only of identical atoms. Compounds with two atoms were binary, those with more than two were tertiary, quaternary, quincuniary, sextenary, septenary, and so on.

John Dalton was born at Eaglesfield, near the town of Cockermouth in Cumbria, England, on September 5 or 6, 1766, the third of six children. His father was a prosperous weaver who also owned 60 acres (24 ha) of land and a stationery shop attached to his weaving workshop. He was a devout member of the Society of Friends (Quakers) and in accordance with Quaker custom did not register John's birth. Consequently, there is some uncertainty surrounding the precise date. John began his education at Pardshow Hall, the Quakers' school in Eaglesfield, where his mathematical skill attracted the attention of Elihu Robinson, another member of the Eaglesfield Quaker community, who tutored him in mathematics, meteorology, and natural science. When the teacher at Pardshow Hall, John Fletcher, retired in 1778, John took his place. In 1781 John joined his elder brother Jonathon in running a Quaker school in Kendal, a much bigger town nearby, where another Quaker, John Gough, befriended him. Gough, the blind son of a Kendal merchant, was interested in a range of scientific subjects. Influenced by Gough, Dalton began writing popular science articles for two magazines, the *Gentlemen's Diary* and *Ladies' Diary.* Gough also encouraged Dalton to start a diary of weather observations. He began a diary in 1787 and made entries in it regularly until his death 57 years later. His diaries recorded a total of more than 200,000 of his observations. He learned much from Gough, and with that knowledge and Gough's encouragement in 1793 Dalton obtained a position teaching mathematics and natural philosophy at the New College, a dissenting academy in Manchester. In 1799 New College moved to York, but Dalton remained in Manchester, earning his living by giving private lessons in mathematics and chemistry.

Dalton's first publication on weather, *Meteorological Observations and Essays,* was published in 1793, but did not sell well. In 1794 he was elected a member of the Manchester Literary and Philosophical Society, where, a few weeks later, he delivered his first paper: "Extraordinary Facts Relating to the Vision of Colours." Both John and his brother were color blind, and this was the first account of how the world appears to someone with this condition. He contributed so much to the understanding of color blindness that it is sometimes known as Daltonism. He also delivered papers on meteorology to the society. The society bought a house for him, which he shared for

Simple

1	2	3	4	5	6	7	8

9	10	11	12	13	14	15	16

17	18	19	20

Binary

21	22	23	24	25

Ternary

26	27	28	29

Quaternary

30	31	32	33

Quinquenary & Sextenary

34	35

Septenary

36	37

many years with the Reverend W. Johns. Dalton was elected the honorary secretary of the society and later its president.

Dalton remained in Manchester, but every year he visited the Lake District in his native Cumbria, and sometimes he traveled to London. He became very famous. In 1804 and again in 1809–10 he delivered a series of lectures at the Royal Institution in London, and in 1822 he was elected a fellow of the Royal Society. He became a corresponding member of the French Academy of Sciences, and in 1830 he was elected one of its eight foreign members.

In 1833 the British government awarded Dalton a pension of £150 a year, and in 1836 increased it to £300 a year. Translated into present-day values, £150 is the equivalent of about £11,200 ($22,400) and £300 is equivalent to £21,700 ($43,400). This was sufficient to allow him to enjoy a higher standard of living than the one he chose. It was also a considerable honor. Then as now the government supported science through funding institutions and through employing scientists directly. It was rare to reward an individual in this way.

Dalton lived simply. His only regular recreation was a game of bowls, which he played every Thursday afternoon. The English game is quite different from bowling. It is played on an absolutely level, closely cropped lawn called a green, or its indoor equivalent, and there are no pins. It is the game legend says Sir Francis Drake was playing on Plymouth Hoe when the Spanish Armada was seen sailing up the Channel.

John Dalton died in Manchester on July 27, 1844. Unfortunately, the house where he lived for so many years, containing records and relics, was destroyed during an air raid in World War II.

(opposite) John Dalton (1766–1844) designed symbols to represent chemical elements, based on the number of atoms the basic particle of each element contained. Key: 1 hydrogen; 2 azote (nitrogen); 3 carbone or charcoal (carbon); 4 oxygen; 5 phosphorus; 6 sulfur; 7 magnesia (magnesium); 8 lime; 9 soda; 10 potash; 11 strontites (strontium); 12 barytes (barium sulfate); 13 iron; 14 zinc; 15 copper; 16 lead; 17 silver; 18 platina (platinum); 19 gold; 20 mercury; 21 water; 22 ammonia; 23 nitrous gas (nitric oxide); 24 olefiant gas (ethylene); 25 carbonic acid (carbon dioxide); 26 nitrous oxide; 27 nitric acid; 28 carbonic acid (carbon dioxide); 29 carburetted hydrogen (methane); 30 oxynitric acid (nitrogen trioxide); 31 sulfuric acid; 32 sulfuretted hydrogen (hydrogen sulfide); 33 alcohol; 34 nitrous acid; 35 acetic acid; 36 nitrate of ammonia; 37 sugar

HENRY CAVENDISH AND THE CONSTANT COMPOSITION OF AIR

In his first scientific paper, published in 1766 in the *Philosophical Transactions of the Royal Society,* Henry Cavendish (1731–1810) demonstrated that inflammable air (hydrogen) is an element. Other scientists had isolated and studied this gas, but its identity was uncertain, because hydrogen is not the only inflammable air. There is also carbon monoxide, released by heating charcoal, and other gases will burn if ignited. Cavendish showed that the gas he had isolated is a distinct substance, different from all other substances.

Cavendish did not call the gas hydrogen. It was Antoine Lavoisier who named it, using the Greek words *hudro,* meaning "water" and *genes,* meaning "born," to reflect its association with water; its German name is *Wasserstoff,* water substance. Henry Cavendish believed the gas was pure phlogiston. He obtained it by reacting metals with sulphuric acid. He reported that with this reaction "their phlogiston flies off, without having its nature changed in the acid, and forms the inflammable air." Cavendish collected what he supposed was phlogiston in a bottle and found that when it mixed with common air and he ignited it with a taper there was a loud noise and almost all of the inflammable air and almost one-fifth of the common air "lost their elasticity and condensed into dew." Joseph Priestley had performed the same experiment and had noted the loud noise, but he ignored the condensation on the inside of the vessel. Cavendish did not. It showed that water is not an element, as Priestley and others supposed, but a compound. He went on to react inflammable air with dephlogisticated air (oxygen), finding that water consisted of one part of inflammable air to 2.02 parts of dephlogisticated air. In modern terms: $2H_2 + O_2 \rightarrow 2H_2O$.

Henry Cavendish was meticulous. His results were usually based on many analyses of samples, and in 1783 he published his findings from 400 analyses of samples of air collected over 60 days. Cavendish reported that common air comprised 20.833 percent dephlogisticated air and 79.167 percent phlogisticated air. His result was very accurate. The actual proportion of oxygen (dephlogisticated air) is 20.95 percent and Cavendish's phlogisticated air consisted of nitrogen (78.08 percent) and argon (0.93 percent), giving a total of 79.01 percent. Argon is an inert gas that forms no compounds, making it extremely difficult to detect. It was not discovered until 1894, by Lord

Rayleigh and William Ramsay (see "Lord Rayleigh, William Ramsay, Noble Bases, and Why the Sky Is Blue" on pages 36–41).

Cavendish's other major contribution to science, also achieved after years of sampling and measuring, was his estimate of the mean density of Earth. He calculated this to be 339 pounds per cubic foot (5,448 kg/m^3); the true density is 343 pounds per cubic foot (5,517 kg/m^3). Cavendish achieved this by using a *torsion balance* to measure the gravitational attraction between two pairs of lead spheres, one pair each weighing 350 pounds (159 kg) and the other pair each weighing 1.61 pounds (0.7 kg). A torsion balance detects very weak forces. It consists of a horizontal bar suspended from a thin fiber that acts as a very weak coiled spring. If weights are suspended from the ends of the bar the fiber will twist, causing the bar to rotate. The amount of rotation is proportional to the force applied to the bar.

Henry Cavendish was born in Nice, France, on October 10, 1731. His family were staying in Nice at the time, but their principal home was in London. The family was aristocratic and very wealthy. Henry's father was Lord Charles Cavendish (1704–83), a scientist who was a fellow of the Royal Society, and the youngest son of William Cavendish, the second duke of Devonshire. Henry's mother was Lady Anne Grey, the daughter of the duke of Kent. Henry's formal education began when, at the age of 11, he attended Dr. Newcome's Academy in Hackney, London. In 1749, when he was 18, he entered St. Peter's College (now called Peterhouse) at the University of Cambridge. He left the university in 1753 without taking the degree examination. This was not unusual at the time, although he may have avoided the examination because his shyness made it impossible for him to endure a direct interrogation by his professors. His father encouraged his scientific interests and introduced him to members and fellows of the Royal Society. Henry was elected a fellow in 1760.

Lord Charles Cavendish taught his son thrift, requiring him to live on an annual allowance of £120, which is probably equivalent to about £17,000 ($34,000) today. He lived with his father in a house in Great Marlborough Street, Soho, and assisted with his father's experiments. After Lord Charles Cavendish's death in 1783 Henry inherited a vast fortune and moved to a substantial house on Clapham Common in south London. He was now a millionaire, but this made little difference to his daily life. He had to pay for the running of the house, including the wages and food of his servants, but most of his expenditure was on books and laboratory equipment. He spent very little on himself and had no time for luxuries. He dressed shabbily, in clothes that had been out of fashion for decades.

Over the years Henry Cavendish built up a substantial library, which other scientists were welcome to use, but he was intensely shy. He kept careful notes of his work, but much of it remained unpublished during his lifetime. Long after his death the Scottish mathematician and physicist James Clerk Maxwell (1831–79) went through Cavendish's papers and found details of several discoveries that had been attributed to others.

Henry Cavendish rarely appeared in public, and he communicated with his housekeeper mainly by leaving notes for her, usually telling her what he wanted for dinner. He had a valet, but female servants had to keep out of his sight. He always attended the dinner the Royal Society Club held prior to its weekly meeting. This was the extent of his social life, and although he enjoyed the company of other scientists he rarely spoke to them. He never married or, so far as anyone knows, enjoyed a close personal relationship with anyone outside his family. He died on February 24, 1810, having dispatched his valet to inform Lord George Cavendish, the son of Henry's cousin, that he was dying. Lord Brougham (1778–1868), who knew him as well as anyone, said later he believed Cavendish wanted to die alone so he could observe and record the process.

He was buried in Derby Cathedral and left his fortune to his family, most of it to Lord George Cavendish. In 1879 the seventh duke of Devonshire, one of Cavendish's descendants, was chancellor of the University of Cambridge. He donated the funds to establish a laboratory to commemorate the family. Scientists working at the Cavendish Laboratory later produced some of the most impressive achievements in physics, a fitting tribute to this brilliant but strange man.

LORD RAYLEIGH, WILLIAM RAMSAY, NOBLE GASES, AND WHY THE SKY IS BLUE

Henry Cavendish had isolated phlogisticated air and by careful measurement had found that it represented 79.167 percent of common air. That was in 1783. In 1892 another eminent scientist, Lord Rayleigh, found a puzzling discrepancy when he compared the density of what he knew as nitrogen obtained from the air and nitrogen that he produced by chemical reactions in the laboratory. It appeared that Cavendish's measurement for the proportion of nitrogen in air was very slightly too high, because Rayleigh found that nitrogen separated from air was invariably 0.5 percent denser than nitrogen obtained by chemical reactions. There were several possible reasons for the discrepancy, and Rayleigh eliminated them one by one, but atmospheric nitrogen remained just a little denser than laboratory nitrogen. This should not be, because chemists and physicists knew that the mass of every element is a multiple of the mass of a hydrogen atom. Nitrogen was breaking the rule. In the end Rayleigh wrote to the journal *Nature* inviting other scientists to help solve the problem. In his note to *Nature* Rayleigh described how he had obtained nitrogen by a method suggested to him by Professor Ramsay, who had noted a similar discrepancy. After *Nature* published the note the two men agreed to work on the problem together, and in 1894 they solved it. Ramsay wrote to Rayleigh that he had isolated from the atmospheric nitrogen gas a heavy component that appeared to be very unreactive. He proposed calling the gas *argon,* the Greek word for "idle," to reflect the fact that it would form no compounds. Rayleigh announced their discovery in a lecture at the Royal Institution in London. In 1904 Lord Rayleigh received the Nobel Prize in physics for the discovery of argon and William Ramsay the Nobel Prize in chemistry for his discovery of argon, krypton, neon, and helium.

William Ramsay was born in Glasgow, Scotland, on October 2, 1852. He studied at the University of Glasgow and from 1870 until 1872 at the University of Tübingen, Germany, where he obtained his doctorate. Returning to Glasgow, in 1872 he obtained a position as a chemistry assistant at Anderson College, and in 1874 he moved to a similar post at Glasgow University. He was appointed principal and professor of chemistry at the University of Bristol in 1880 and in 1887 he became professor of inorganic chemistry at University College, London. He remained in this post until his retirement in 1913.

Ramsay was knighted in 1902 and was an honorary member of many European scientific academies. After retiring he moved to High Wycombe, north of London, where he died on July 23, 1916.

Sir William Ramsay was a chemist. He isolated helium from a uranium compound in 1895. In 1898 he and his colleagues obtained 26 pints (15 liters) of what they thought was pure argon in liquid form. Argon boils at -301°F (-185°C), but when they allowed it to boil a more volatile gas was released before the argon itself boiled. They called this gas *neon,* Greek for "new." Neon boils at the lower temperature of -410.89°F (-246.05°C). Then the argon boiled away, leaving two heavier gases. They called one of these *krypton* (boiling point -242.14°F, -152.3°C) from the Greek *krupton* meaning "hidden" and the other *xenon* (boiling point -160.78°F, -107.1°C), from *xenos* meaning "strange." Argon, neon, krypton, and xenon are known as noble gases, because at first they were thought to form no true, stable compounds. Xenon does form a few compounds, however, and neon forms compounds with fluorine.

Lord Rayleigh was a physicist with a particular interest in waves. He studied light waves, sound waves, ocean waves, and seismic waves. As part of his research, he also studied the effect of the atmosphere on light waves passing through it. Rayleigh found that very small atmospheric particles, with a size comparable to the wavelength of light, scatter light in all directions. The way this happens varies according to the size of the particles and the wavelength of the light. Air molecules—nitrogen and oxygen—scatter short wavelengths more than longer wavelengths. In descending order of wavelength (starting with the longest), the spectrum of white light consists of red, orange, yellow, green, blue, indigo, and violet. Violet and indigo light, with the shortest wavelengths, are scattered to such an extent that they disappear entirely. The next wavelength is blue. It is scattered in all directions. This means that wherever an observer looks into a cloudless sky, except close to the Sun where the brightness makes it impossible to distinguish colors, the light reaching the eye is predominantly blue. This type of *scattering* is known as Rayleigh scattering. There is a second type of scattering, discovered by the German physicist Gustav Mie (1868–1957) and called Mie scattering. It is caused by particles larger than the wavelength of light. These scatter all wavelengths of light in a forward direction. That is why the sky appears white when it contains a large amount of dust.

Rayleigh scattering also explains the colors of sunset and dawn. At these times the Sun is very low in the sky. Consequently sunlight travels through a much greater thickness of air to reach the ground as seen in the illustration. The blue light and most of the green are scattered away almost completely, allowing the longer wavelengths of yellow, orange, and red to predominate.

Lord Rayleigh was born John William Strutt on November 12, 1842, at Maldon, Essex. He was the son of the second baron Rayleigh and Clara Elizabeth La Touche Vicars. His health was frail, and illness interrupted his education many times. At age 10 he was sent to Eton College, but spent most of his short time there in the school sanatorium. He then attended a private school in Wimbledon, followed by a short spell at Harrow School. Finally, in 1857, he went to Torquay, in Devon, to stay with the

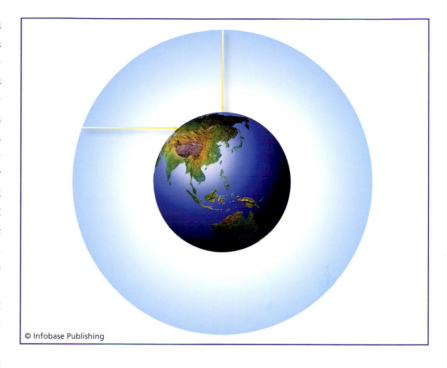

© Infobase Publishing

When the Sun is high in the sky its light travels through a thinner layer of the atmosphere than is the case when it is low in the sky and its light arrives obliquely.

Reverend George Townsend Warner, who taught pupils privately. After spending four years at Torquay, in 1861 Strutt entered Trinity College, University of Cambridge, to study mathematics. He graduated in 1865, and in 1866 obtained a fellowship at Trinity College, which he held until his marriage in 1871 to Evelyn Balfour.

An attack of rheumatic fever in 1872 compelled him to spend the winter in Egypt and Greece. He resumed his scientific work as soon as he recovered. Then, in 1873 his father died, and Strutt acceded to the title, becoming the third baron Rayleigh (Lord Rayleigh). He went to live in the family home, Terling Place at Witham, Essex, and had to devote some of his time to managing the 7,000-acre (2,800-ha) estate that he had inherited. He proved a skilled and progressive manager, but in 1876 he handed over the entire running of the estate to his younger brother. This allowed him to concentrate fully on his science.

In 1879 Lord Rayleigh was appointed professor of experimental physics at Cambridge and head of the Cavendish Laboratory, but he left Cambridge in 1884 to continue his experimental work at Terling Place. He was professor of natural philosophy at the Royal Institution from 1887 until 1905. Eventually he became chancellor of Cambridge

University. He was elected a fellow of the Royal Society in 1873 and was its secretary from 1885 until 1896 and its president from 1905 until 1908. Lord Rayleigh received the Order of Merit in 1902 and in 1905 he was made a Privy Councillor. The Order of Merit (OM after the holder's name) is a British Order of Chivalry founded by Edward VII in 1902. Membership is awarded by the British sovereign to people of outstanding intellectual achievement—scientists, artists, writers, scholars, and politicians—and the order can have only 24 members at any one time. The Privy Council is a committee of the British sovereign's closest advisers. Lord Rayleigh died at his home on June 30, 1919.

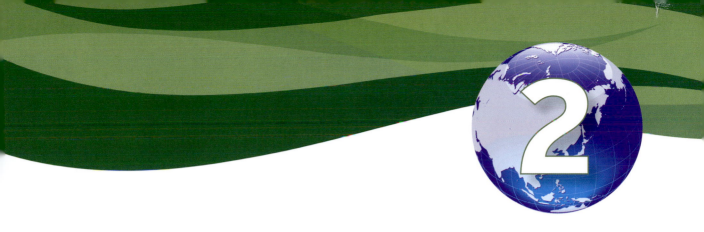

Measuring the Air

Scholars in the ancient world learned to measure distances and angles. They were able to survey land and calculate areas, and their astronomers made detailed studies of the night sky. They produced engineers and architects who built structures that are still among the wonders of the world. But they could not weigh or measure the atmosphere. Their understanding of weather extended no further than associating effects with obvious causes.

It was not until the end of the 16th century that anyone even attempted to make a thermometer or barometer. This is less extraordinary than it may seem, for it was not until then that physicists began to appreciate the value as well as the feasibility of quantifying natural phenomena. It was no longer sufficient to say, for example, that yesterday the air felt warmer than it feels today. Students of the atmosphere needed to know how much warmer it was, and as they began to make measurements more questions arose. What is temperature and is it the same thing as heat? (No, it is not.) If air has weight, how is it that its weight changes from one day to another?

This chapter describes some of the early efforts to measure the principal qualities of air—its temperature and pressure. It begins with the first thermometer, made by Galileo, the most renowned of all European scientists, and explains why it did not work. Later, the chapter describes the thermometers that did work and the scales used to calibrate them, associated with two household names: Fahrenheit and Celsius.

Galileo also posed the question that required the invention of a new instrument. He died before it was made, but that instrument not only answered the question but also discovered that the atmosphere exerts a pressure, which is changeable. It was the barometer. There is a barometer on the wall in many homes today. The chapter describes how the most common type of modern domestic barometer was designed in the 17th century.

GALILEO AND THE THERMOMETER THAT FAILED

Aristotle (384–322 B.C.E.) taught his students to observe the world around them, but his explanations of natural phenomena were based entirely on rational argument. He used reason as his only tool. His abstract approach dominated European thought until the second half of the 16th century, when an entirely different approach began to emerge, promoted by chemists and physicists who believed in quantifying phenomena. Quantification yielded values expressed in numbers that could be processed mathematically to produce numerical solutions that could be verified. It was the beginning of modern science, and its most famous exponent was Galileo Galilei (1564–1642), who is usually known simply by his first name, Galileo.

Galileo designed experiments that produced results he could measure and then analyze. At the same time, he recognized that no measurement could ever be absolutely precise and for that reason experimental results would usually differ slightly from the results predicted by mathematical calculation. Stephen Hawking (born 1942), the Lucasian Professor of Mathematics at the University of Cambridge, wrote in *A Brief History of Time* (London: Bantam Press, 1988) that: "Galileo, perhaps more than any other single person, was responsible for the birth of modern science." And in *Ideas and Opinions* (London: Crown Publishers, 1954) Albert Einstein (1879–1955) wrote: "Propositions arrived at by purely logical means are completely empty as regards reality. Because Galileo realized this, and particularly because he drummed it into the scientific world, he is the father of modern physics—indeed, of modern science altogether."

Galileo's reputation rests principally on three achievements. He was the first person to use a telescope to study the night sky, and his observations provided evidence supporting the conclusion of Nicolaus Copernicus (1473–1543) that the Sun and not the Earth lies at the center of the universe and that the Earth orbits the Sun. His studies of gravity and motion established the principles that Isaac Newton (1642–1727) later formalized as the laws of motion. His most important achievement, however, was his application of mathematics to the study of natural phenomena.

Reliable instruments were essential to this enterprise, because they made it possible to compare measurements taken at different times and in different places. Galileo famously made his own telescopes—and turned this into a profitable sideline making spyglasses for mariners. The telescope was invented in the Netherlands, and Galileo worked from descriptions that reached him. In about 1593 he also made the first thermometer.

Strictly speaking, Galileo's instrument was not a thermometer, because it had no scale that would allow differences in temperature to be quantified. It was a *thermoscope* that made changes in temperature visible. The Greek mathematician and engineer Hero of Alexandria (ca. 10–70 C.E.) had made devices that worked by the expansion and contraction of air as its temperature changed, and Galileo sought to exploit the same principle, so the instrument he made was an air thermoscope.

The following illustration shows how the air thermoscope worked. The device consisted of a long, narrow glass tube, open at one end and with a bulb at the other. The bulb was heated, causing the air inside it to expand and, while it was still hot, the open end of the tube was placed below the surface of a colored liquid in a flask and the tube secured in an upright position. As the bulb cooled the air inside it contracted and some of the colored liquid was drawn upward. Thereafter, the level of colored liquid in the tube would rise or fall as the temperature of the air in the bulb changed.

At least, that is what should have happened, but unfortunately there was a problem, and the air thermoscope proved highly unreliable. The top of the flask was not sealed, and what Galileo did not know was that atmospheric pressure exerted a force on the surface of the colored liquid. Changes in pressure made the liquid rise or fall in the tube independently of the temperature, so the thermoscope was simultaneously recording changes in both pressure and temperature, with the inevitable result that neither was accurate because there was no way to separate the two effects. Nevertheless, Galileo's air thermoscope was a prototype on which other instrument makers could improve.

As air expands and contracts in the bulb at the top, the thermoscope pushes the colored liquid down or draws it up.

© Infobase Publishing

Galileo Galilei was born in Pisa, in the Grand Duchy of Tuscany, on February 15, 1564. He was the eldest of the seven children of Vincenzo Galilei (1520–91), a famous lutenist and expert in music theory, and Giulia Ammanati (1538–1620). He received his first lessons from a private tutor. In 1574 the family moved to Vallombrosa, near Florence, and Galileo attended lessons at a nearby monastery. In 1581 he enrolled at the University of Pisa to study medicine. The family could not afford the expense, and in 1585 he returned home without having obtained a degree. While at the university he had developed a keen interest in mathematics and physics, and by the time he left he had started to study them.

Back in Florence, Galileo obtained a post as lecturer in mathematics and science at the Florentine Academy. In 1589 he was appointed professor of mathematics at the University of Pisa. The appointment was an honor, but the post was poorly paid, and in 1591 Galileo's father died, leaving Galileo to care for his 16-year-old brother Michelagnolo (sometimes spelled Michelangelo, 1575–1631). In 1592 Galileo applied successfully for the better-paid position of professor of mathematics at the University of Padua. He and Michelagnolo moved to Padua, where Galileo remained for 18 years. It is where he did his best work.

Galileo was argumentative, and he could be sarcastic. He had little time for those who disagreed with him, but he was forging an entirely new approach to physics founded on his deep conviction that natural phenomena can be described mathematically, and that careful experiment and observation will produce data that validate the mathematical description.

He declared his support for the Copernican view of the universe against the traditional model based on the teachings of the Greek astronomer and mathematician Ptolemy (Claudius Ptolemaeus, 83–161 C.E.) in the form of a dialogue in his book *Dialogo Sopra I Due Massimi Sistemi del Mondo* (Dialogue concerning the two chief world systems). It was published in Florence in 1632, with the permission of the Inquisition, but complaints and further investigation brought him into conflict with the church, and in 1633 he faced trial before the Inquisition in Rome. He was found guilty of "grave suspicion of heresy." His book was banned (the ban was lifted in 1822), and Galileo was sentenced to remain for the rest of his life at his villa at Arcetri, near Florence, under a form of house arrest. The ruling did not prevent him from continuing his work, but his eyesight was failing, and in 1637 he became blind. With the help of a team of assistants he designed a pendulum-driven clock that was built in 1656 by Christiaan Huygens (1629–95). Galileo was still dictating to his secretaries when he fell ill with a fever late in 1641. He died at Arcetri on January 8, 1642.

FERDINANDO OF TUSCANY, INVENTOR OF INSTRUMENTS

Galileo was a citizen of Tuscany, which was a grand duchy ruled by the Medici family, and it was Grand Duke Ferdinando I (1549–1609) who appointed Galileo to the professorship of mathematics at Pisa. Ferdinando was interested in science, and Galileo tutored his eldest son, later Cosimo II (1590–1620). When Cosimo died, his son was only 10 years old. The boy inherited the title and became Ferdinando II (1610–70).

All of the Medici rulers were enthusiastic supporters of the arts and sciences, but Ferdinando II was the foremost Medici scientist, with a particular interest in meteorology. Ferdinando knew Galileo and consistently supported his work, and he certainly knew about the air thermoscope and its unreliability. He also knew why it was unreliable, and in 1654 he solved the problem by designing a thermometer containing alcohol, filling the bulb and part of the tube with liquid, then melting the tip of the tube to seal it from the outside air, so it was not affected by changes in atmospheric pressure. In 1612 the Italian physician Santorio Santorio (1561–1636) had also made an air thermometer that was no more accurate than Galileo's, but Santorio had devised a graduated numerical scale to measure temperature changes. Santorio used his thermometer to measure the temperature of patients, and he was the first physician to do so. Ferdinando adapted Santorio's scale and produced an instrument that was well on the way to being an accurate, reliable thermometer. The first scientists to use thermometers containing mercury were members of the Accademia del Cimento (Academy of Experiment) in Florence.

The Accademia del Cimento was founded in 1657 by students of Galileo and financed by Ferdinando and his brother Prince Leopoldo (1617–75). The institution survived for only 10 years, but it was one of the world's first scientific academies (see sidebar, page 47). Members of the Accademia were especially interested in atmospheric science, so developing a practical thermometer was central to their endeavors. In 1694 Carlo Renaldini (1615–98) proposed a solution to a serious calibration problem that had emerged, raising the possibility of an improvement on the Santorio scale. Carlo Renaldini (sometimes spelled Rinaldini) was an aristocrat from Ancona. He was principally a mathematician and was professor of philosophy at the University of Pisa.

SCIENTIFIC ACADEMIES

During the 15th century educated Europeans came to regard ancient Greece and Rome as the source of their culture and learning. The works of Greek philosophers became widely available in Latin translations, often having being translated first into Arabic and from Arabic into Latin. Latin was then the language of European scholarship and international communication; far fewer people understood Greek. Growing interest in Greek philosophy led to attempts to revive the schools that once flourished in Athens, especially the academy of Plato. The early academies aimed to be republics of letters where intellectuals could study, discuss, and promote the values of the philosophers on whom they modeled themselves. The most famous of these was called the Platonic Academy. It was founded in Florence in 1459 under the patronage of Cosimo de' Medici expressly for the purpose of studying the works of Plato and disseminating Latin translations of them.

By the 17th century intellectual interests had widened. Although scholars continued to study the languages, literature, arts, and philosophies of the ancient world, an increasing number were more concerned with the study of nature. They began to form academies dedicated to scientific pursuits. In 1609 Galileo Galilei (1564–1642) joined the first of these scientific academies, the Accademia dei Lincei, which lasted from 1603 until 1630 in Florence. Its name, which translated to Academy of Lynxes, reflected the expectation that its members would study nature with the keen eyesight of a lynx. The Accademia del Cimento, also in Florence, lasted from 1657 until 1667.

Scientists in other countries then began forming their own academies. In 1660 British scientists secured the patronage of Charles II to establish the Royal Society of London, with Robert Boyle (1627–91) as one of its founding members, and Robert Hooke (1635–1703), Isaac Newton (1642–1727), and John Locke (1632–1704) joining in the following years. Members of the Royal Society corresponded with colleagues in other countries. In France, the Académie Royale des Sciences, founded in 1666, paid leading scientists and provided accommodation for them.

In the years that followed, the examples of the London and French academies prompted the formation of many more. In Bologna, the Istituto delle Scienze was founded in 1711 and created professorships in subjects not being covered adequately by the universities. Istituto professors gave lessons at times that did not conflict with university lectures so students could attend both. Peter I (the Great) founded the Russian Academy of Sciences in St. Petersburg in 1724. In the United States, President Abraham Lincoln signed the document establishing the National Academy of Sciences on March 3, 1863.

Today most nations have an academy of sciences. Membership or fellowship of the academies is through election by the existing members on the basis of significant contributions to some aspect of science. Specialist committees of academy members then offer advice to governments and the public and publish original research. For example, the National Academy of Sciences publishes the *Proceedings of the National Academy of Sciences* and the Royal Society publishes *Philosophical Transactions of the Royal Society.*

It was widely agreed that the temperature scale should extend from the freezing point of water to its boiling point and at first it seemed that constructing the scale would involve nothing more complicated than marking those two points and dividing the distance between them into equal units. When this was tried, however, the instrument makers found that water does not expand at a constant rate as its temperature rises, and they assumed the same would be true of any other liquid. Renaldini's solution was to relate the scale directly to the behavior of water by mixing freezing and boiling water in varying proportions. First, the thermometer bulb would be immersed in freezing then boiling water and the two positions of the thermometer liquid marked. Next, equal amounts of boiling and freezing water would be mixed and the thermometer placed in the mixture. This would indicate the halfway position between freezing and boiling, which would be marked as the center point on the scale. Then the intermediate points would be determined in the same way. Mixing 20 parts of freezing water with 80 parts of boiling water would show the position one-fifth of the distance from the freezing point, and so on. It was an ingenious idea, but proved difficult to put into practice.

Ferdinando de' Medici was a popular ruler, with a mild temperament and friendly disposition. He encouraged trade, signing several trade treaties and sponsoring the development of the harbor at the Adriatic port of Leghorn (Livorno). During the trial of Galileo he tried hard to persuade the Inquisition that Galileo was innocent and should be allowed to continue with his work. When that failed, and Galileo was sentenced to house arrest and forbidden to publish, Ferdinando continued to try to have the conviction overturned or the sentence reduced. In the 1640s Ferdinando began to encourage scientific experimentation at his court, and it was during this period that he began working on his thermometer. He also invented a *hygrometer* to measure atmospheric *humidity* (see "Guillaume Amontons and the Hygrometer" on pages 68–73 and the sidebar "Humidity" on page 69), as well as a hydrometer to measure the density of liquids. In 1644 one of the court experiments led to the construction of the first artificial incubator, located in the citrus greenhouses at the Boboli Garden in Florence, in which a thermometer regulated the temperature beneath a brooding hen.

Ferdinando was unable to prevent the steady decline in Tuscan power, and he could not sway the Inquisition in the case of Galileo.

In 1634 he married his cousin Vittoria della Rovere (1622–94). They had two children, Cosimo (1642–1723), who succeeded Ferdinando as Grand Duke Cosimo III de' Medici, and Francesco Maria (1660–1711), who became a cardinal and was governor of Siena from 1683 until his death. Cosimo III had two sons and one daughter, but neither of his sons produced male heirs, and the Medici family ended with the death of Gian-Gastone, Ferdinando's grandson, in 1737.

EVANGELISTA TORRICELLI AND THE FIRST BAROMETER

In 1641 Galileo was old and blind, but his mind remained as keen as ever, and as active. He was no longer able to write or conduct experiments, but his assistants and secretaries pursued his ideas and wrote from his dictation.

One of the questions puzzling scientists at the time, including Galileo, was why a piston is able to draw water up a cylinder. When the open end of a cylinder is placed in water and a close-fitting piston is drawn upward through the cylinder, water follows it. But why? The prevailing explanation was based on the belief that vacuums cannot exist, an idea that began with Aristotle. In about 350 B.C.E. Aristotle was engaged in a debate between those who believed all matter consisted of atoms and voids and, opposing them, those who believed no such thing as a void could exist and, therefore, that apparently empty space was filled with some kind of substance. Aristotle maintained that the speed of a moving body is limited by friction. If a body moved through a void, containing no matter of any kind, there would be no friction, and the body could accelerate to an infinite speed. He found it impossible to accept the idea of infinite speed and therefore denied the possibility of a vacuum with the famous saying: "Nature abhors a vacuum."

Galileo accepted this, and therefore the general explanation for the piston effect. This held that withdrawing the piston creates a vacuum in the cylinder behind the piston. Nature abhors a vacuum, so water rises to fill it. This was quite satisfactory for pistons moving through short cylinders, but not for the long cylinders, with pistons operated by systems of levers, which were used to pump water. In a long cylinder water will rise no more than about 33 feet (10 m). Beyond that the piston would continue to move upward, but the water would rise no higher. Galileo thought it possible that although nature abhors a vacuum in general, perhaps it will tolerate one under certain circumstances. This was a weak, unsatisfactory explanation and Galileo knew it, so he passed the question to one of his assistants for further investigation.

The assistant Galileo chose was Evangelista Torricelli (1608–47). Torricelli rejected the Aristotelian idea of nature abhorring a vacuum. Instead he thought about the properties of air. Physicists at the time believed air possesses a property of levity, which makes it rise, but Torricelli suggested a very radical alternative: What if air has weight? If that were the case, the weight of air would press down on the surface of the water outside the cylinder. As the piston was withdrawn leaving an empty space, the pressure outside the cylinder would push water higher inside the cylinder, where there was no overlying weight of air to press down on it. If he were right, that would explain why water rose in the cylinder, but not why it rose for only a limited distance. Torricelli suggested that might be explained if the weight of the air was sufficient to push water upward for only that distance. In that case, withdrawing the piston farther up the cylinder would have no effect on the level of water, because the height of water inside the cylinder was determined wholly by the weight of the air pressing on the water outside the cylinder. That is the hypothesis Torricelli set out to test in 1643.

Torricelli realized that his experiment would be easier if he used a liquid that was denser than water, because if he used water he would have to manipulate a tube more than 33 feet (10 m) long. He chose mercury, which is 13.6 times denser than water. He poured a quantity of mercury into an open-topped dish and completely filled a tube 4 feet (1.2 m) long that was open at one end. Covering the open end to prevent the mercury falling out, he placed it below the surface of the mercury in the dish, then raised the tube and secured it in an upright position. When he did so, some of the mercury flowed from the tube into the dish, but not all of it, leaving the surface of the mercury in the tube at a higher level than that of the mercury in the dish. No piston had been withdrawn to pull mercury upward or to create a vacuum that mercury rushed to fill. The only force able to support the mercury in the tube was the weight of air pressing down on the mercury in the dish. He had proved that air has weight, a discovery with implications Blaise Pascal (1623–62) would explore a few years later (see "Blaise Pascal and the Change of Pressure with Height" on pages 106–109). The following illustration shows how Torricelli's experimental apparatus worked.

Torricelli had also created the first artificial vacuum in the air above the mercury in the tube. Apart from a very small amount of mercury vapor, that part of the tube was empty. Torricelli had demonstrated that the proposition "nature abhors a vacuum" is false, and a vacuum produced in this way is still known as a *Torricelli vacuum.*

When Torricelli measured the height of the mercury in the tube he found it to be approximately 30 inches (760 mm). He did not dismantle his apparatus immediately, and visiting it at various times in the succeeding days he noted that the height of mercury in the tube varied slightly. He measured the changes and concluded, correctly, that they were due to changes in the weight of air pressing downward. Torricelli had discovered air pressure and had invented the

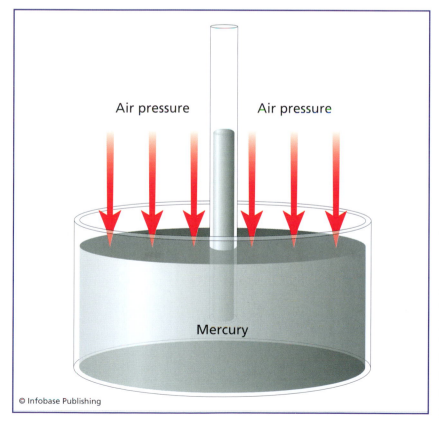

Air pressure   Air pressure

Mercury

© Infobase Publishing

Torricelli's barometer. Air pressure exerts a force on the mercury in the reservoir, but a vacuum fills the tube above the level of the mercury, so no air pressure acts on the mercury in the tube. Mercury from the reservoir rises up the tube to a level proportional to the magnitude of the air pressure.

barometer. In a letter he wrote in 1644, Torricelli remarked that: "We live submerged at the bottom of an ocean of elementary air, which is known by incontestable experiments to have weight."

Evangelista Torricelli was born on October 15, 1608, at Faenza, near Ravenna. His father died while he was young, and Evangelista was brought up by his uncle, who was a monk belonging to a branch of the Benedictine order. In 1624 his uncle enrolled Evangelista at a Jesuit college, where he studied mathematics and philosophy, and in 1627 he sent the boy to the Collegio della Sapienza in Rome, where the professor of mathematics was another Benedictine, Benedetto Castelli.

Castelli had been a student of Galileo, and Torricelli read works by Galileo, including his *Dialogo Sopra I Due Massimi Sistemi del Mondo* (Dialogue concerning the two chief world systems), which had not yet been proscribed (see "Galileo and the Thermometer That Failed" on pages 42–45). He wrote a letter to Galileo praising his work and asserting his conversion to the Copernican view. He also wrote a treatise on the paths of projectiles, developing some of Galileo's ideas on mechanics. Castelli, much impressed by the quality of his student's work, sent the papers to Galileo with a covering letter. Galileo read them and was also impressed and immediately invited Torricelli to visit him at Arcetri, where by that time he was confined. Torricelli did not leave Rome at once, and by the time he reached Arcetri in 1642 Galileo had only three months left to live. Galileo accepted Torricelli as one of his assistants and set him the problem of the water in the cylinder. Unfortunately, the great man died before the solution was found.

On January 8, 1642, Ferdinando II invited Torricelli to take Galileo's place as mathematician to the Tuscan court, and he was also appointed professor of mathematics at the Florentine Academy. Torricelli died in Florence from typhoid fever on October 25, 1647.

ROBERT HOOKE AND THE WHEEL BAROMETER

Evangelista Torricelli observed that the height of the mercury in his vertical tube changed from time to time, and he deduced that this was because the weight of the air was slightly variable. He did not link these changes in the weight of air to changes in the weather. The first person to make that connection was the English mathematician,

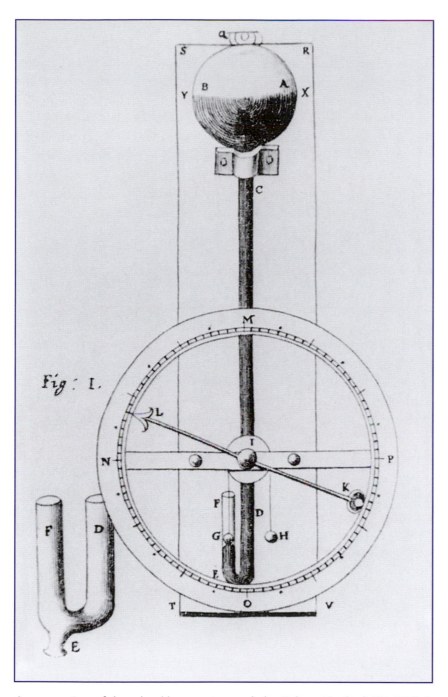

An engraving of the wheel barometer made by Robert Hooke (1635–1703). It converted the vertical movement of mercury in a tube to the rotary motion of a needle. *(Science Photo Library)*

physicist, and instrument maker Robert Hooke (1635–1703). Hooke noticed that storms often followed a sharp fall in the mercury level.

Hooke was one of the most ingenious inventors and experimenters the world has ever seen. Some people have described him as the "English Leonardo," comparing him to Leonardo da Vinci. Hooke proposed the relationship now known as Hooke's law, which states that the stress placed on an elastic body is proportional to the strain it produces. He invented an escapement mechanism for watches, claiming it was the first, although Christiaan Huygens also claimed priority, leading to a bitter dispute. Hooke was probably in the right. He greatly improved the design of microscopes, and his book *Micrographia*, published in 1665, contained the first drawings, by Hooke, of observations through a microscope. He was the first person to observe plant cells, and it was he who named them cells, because they reminded him of monks' cells. He designed flying machines, and he showed that long ago Britain lay beneath the sea. His list of achievements is long.

The Torricelli barometer recorded changes in pressure by small vertical movements of the level of mercury in its tube. Robert Hooke thought the instrument would be much easier to use if the changes could be magnified and shown by movements of a needle on a dial, and he devised a way to convert the vertical movement of the mercury to the rotary movement of a needle. His own illustration shown below of what came to be known as his wheel barometer demonstrates how the instrument worked. He turned the Torricelli barometer upside down, with the mercury reservoir at the top and open to the air, so air pressure was free to act on the surface of the mercury. The lower end of the tube formed a U-shaped bend. The tube was open and a float rested on top of the mercury. The float was linked to a wheel, and the needle was attached to the wheel. As the mercury in the U-bend rose and fell, the linkage mechanism made the needle turn, and its point showed the amount of air pressure on a calibrated dial. The drawing shows the barometer stripped of its housing. It would have been enclosed in a wooden case that concealed the reservoir and tube, leaving only the needle and dial visible. This type of barometer is sometimes called a banjo barometer because of its shape.

Having recognized the connection between air pressure and weather conditions, Hooke sought to make his wheel barometer still more user-friendly by marking around the dial the likely effects of

THE ANEROID BAROMETER

A mercury barometer must have a reservoir and a tube that is rather more than 30 inches (76.2 cm) long, because that is the height mercury will reach at average sea-level pressure. It is clearly impractical to use any alternative liquid, because it would be less dense than mercury and the tube would need to be proportionately longer. Alcohol works satisfactorily in a thermometer, but an alcohol barometer would be almost 150 feet (46 m) tall! The length of the tube determines the size of a mercury barometer and although the pressure may be displayed by a needle and dial, the barometer must still have a long tube.

An alternative, based on an entirely different principle, was invented in 1843 by the French engineer Lucien Vidie (1805–66). Although less accurate than the mercury barometer, it is small and easily portable, and it is now the most popular design for household barometers. It uses no liquid, so it is called aneroid from the Greek words *a*, meaning "no," and *neros*, meaning "wet."

(continues)

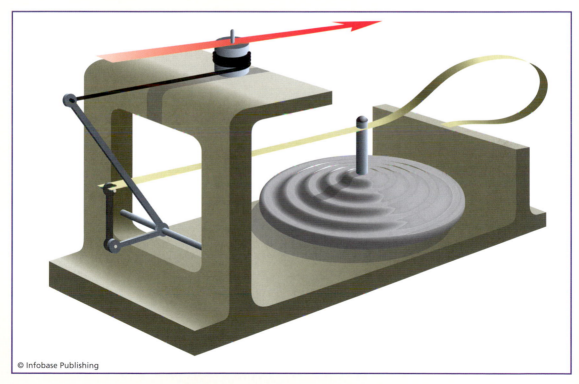

© Infobase Publishing

Invented in 1843, the aneroid barometer contains a corrugated metal box that expands and contracts with variations in air pressure. A spring and levers transmit movements of the box surface to a needle that moves over a dial.

(continued)

An aneroid barometer contains a vacuum box consisting of two corrugated sheets of metal fastened together to enclose a space from which the air is evacuated. Changes in air pressure make the two sheets move closer together or farther apart. These movements are transmitted to a spring, which magnifies them, and then to a system of levers linked to a needle that moves against a dial. The illustration (overleaf) shows the mechanism inside a barometer of this type.

a change in pressure. He marked his dial with the words *change, rain, much rain, stormy, fair, set fair, and very dry.* To this day, many household barometers still use these labels or ones very like them. The labels are often seen on *aneroid barometers,* which were not invented until the 19th century (see the following sidebar).

Hooke was keenly interested in the weather and is considered one of the founders of the science of meteorology. In 1450 the Italian writer, artist, and architect Leon Battista Alberti (1404–72) made the first *anemometer,* a device to measure wind speed. It consisted of a disk that hung vertically. In still air the disk hung directly downward, but when there was a wind the disk was deflected. The angle by which it was deflected from the vertical was proportional to the wind speed. In 1667 Hooke improved on this device. *The Posthumous Works of Robert Hooke, M.D.S.R.S.,* edited by R. Waller and published in 1705, contains an account of a presentation Hooke made to the Royal Society on November 14, 1683: "Mr. Hooke shew'd an instrument to measure the velocity of the air or wind and find the strength thereof which was by four vanes put upon an axis and made very light and easy for motion; and the vanes so contrived as that they could be set to what slope should be desired." The Alberti anemometer was the ancestor of the modern swinging-plate anemometer. Hooke's design led to the rotating-cups anemometer. The following illustration shows both these types in their modern forms.

In 1663 Sir Christopher Wren (1632–1723) designed a weather clock that would record weather conditions automatically. Hooke collaborated with Wren to make the first working model, which they called the weather wiser, with trip hammers that recorded pressure, temperature, rainfall, humidity, wind speed, and wind direction as marks on a rotating drum. The two scientists also collaborated to make the first tipping-buckets rain gauge (see "Measuring Rainfall" on pages 81–83). Hooke believed it might be possible to forecast the weather if meteorological conditions were recorded every day at a network of stations.

Robert Hooke was born on July 18, 1635, at Freshwater, on the Isle of Wight, an island close to the southern coast of England. His father, John Hooke, was an ordained minister employed as the assistant to the parish priest. Robert's health was frail and although John wished all four of his sons to enter the ministry, severe headaches prevented Robert from studying as intensely as this required. Too ill to attend lessons, he spent much of his time alone, amusing himself by making

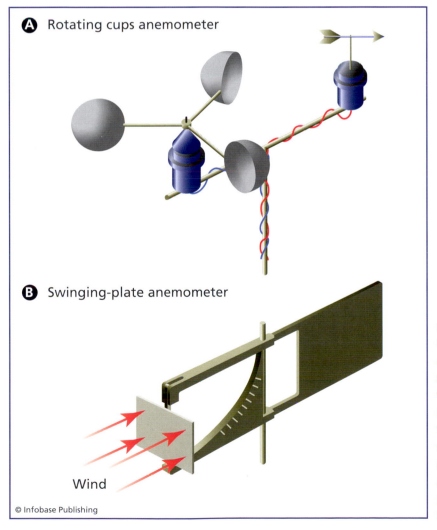

A Rotating cups anemometer

B Swinging-plate anemometer

Wind

© Infobase Publishing

An anemometer measures wind speed and direction. A. The rotating-cups anemometer spins on a vertical axis, the speed of rotation being proportional to the wind speed. B. The swinging-plate anemometer measures wind force by the amount of deflection of a plate held at right angles to the wind.

mechanical toys. His father died when Robert was 13, and he went to London, where he embarked on an apprenticeship to the portrait artist Sir Peter Lely (1618–80), which cost him the £100 he had inherited from his father. Robert abandoned the apprenticeship and enrolled at Westminster School, where he studied Latin and Greek and where he excelled in mathematics. It took him just one week to master the first six books of the geometry of Euclid. Hooke must have had a pleasant voice, because in 1653 he secured a place as a chorister at Christ Church College, at the University of Oxford—although there was probably little call on his singing talents, because there would have been no choir during the Commonwealth period (1649–60). After a time his status changed and he became a servitor, which was an undergraduate student who received financial help from college funds, in return for which he was required to perform certain menial tasks. He did not sit for the examination for a degree.

The skill he had developed making mechanical toys now stood him in good stead, because he was able to earn useful amounts of money by designing scientific instruments and improving on existing instruments. He would take his prototypes to the professional instrument makers, who paid him for his designs and modifications.

While he was at Oxford, Hooke met several of the leading scientists of the day and found employment, first as an assistant to the anatomist Dr. Thomas Willis (1621–75) and in 1658 to the chemist and physicist Robert Boyle (1627–91; see "Robert Boyle, Edmé Mariotte, and Their Law" on pages 101–104). Boyle was wealthy and paid Hooke well, and the two became lifelong friends. Hooke ceased to work for Boyle in 1662, but Boyle continued to pay him until 1664, when Hooke was in a better financial position. Hooke's first task for Boyle was to make a pneumatic pump that could evacuate the air from a container in which Boyle would place instruments and animals to observe the effect. Hooke made several pumps, and in 1671 he tested one on himself by sitting inside a chamber while some of the air was evacuated. He suffered pain in his ears and deafness—and may well have been the first person to experience and document high-altitude sickness.

The Oxford friends began to go their separate ways in 1659, and Boyle and Hooke moved to London, where in 1660 they joined with others to form a scientific society. In 1662 their society became the Royal Society of London (see "Scientific Academies" on page 47), and Robert Hooke was appointed curator of experiments, for which he was paid £30 a year—about £2,900 ($3,800) in today's money—and provided with living accommodation in Gresham College, where the society held its meetings. Gresham College is an institution with the Lord Mayor of London as its president that employs professors to give free public lectures. Hooke's job was to demonstrate experiments at the weekly meetings of the society. He was elected a fellow of the society in 1663 and in the same year Oxford awarded him an M.A. degree. Also in 1663 the society made him a lecturer in mechanics, at the higher salary of £50 a year—about £5,500 ($11,000) in today's money—and in 1665 he was made professor of geometry at Gresham College, at an annual salary of £50. He remained in these two positions for the rest of his life. He was secretary to the Royal Society from 1677 until 1683.

As well as being a skilled instrument maker, experimenter, and mathematician, Robert Hooke was also a talented and successful architect. Following the Great Fire of London in 1666, Hooke was the official surveyor and examined about half of all the building plots in the city. In 1675 he designed the Bethlem Royal Hospital at Moorfields, outside the city boundary. Bethlem is the world's oldest psychiatric hospital (it still exists) and for a long time was known as Bedlam. Hooke designed the Royal College of Physicians, built in 1679, several large houses outside London, and he collaborated with Sir Christopher Wren in designing the Royal Observatory at Greenwich, the Monument (to the Great Fire) in the City of London, and St. Paul's Cathedral. It was Hooke who designed the method for constructing the dome of St. Paul's. He proposed redesigning the streets of London on a grid pattern, with wide boulevards and arterial routes allowing speedy access to the city center.

Robert Hooke never married. His health began to deteriorate in about 1700. He suffered from swollen legs and insomnia. Always thin, he became emaciated, and he was blind. He died at Gresham College in London on March 3, 1703.

DANIEL FAHRENHEIT AND HIS THERMOMETER

Grand Duke Ferdinando II made his first thermometer in 1641 and an improved version in 1654, but still there were no units that would allow scientists to ascribe a numerical value to changes in temperature. The only information a thermometer would convey is that air or a liquid was warmer or cooler than it had been earlier. There was no way of knowing by how much the temperature had changed.

Thermometers needed scales, and in 1701 the English mathematician and physicist Sir Isaac Newton (1642–1727) pointed out that a temperature scale used on a thermometer would need to be calibrated by making a convenient number of equal divisions between two fixed points, called *fiducial points* (see sidebar below). Newton suggested using the *freezing* temperature of water and average body temperature as the fiducial points.

The Danish astronomer and instrument maker Ole Christensen Rømer (1644–1710) met Newton on a visit he made to England in 1679, before Newton proposed his fiducial points. Perhaps the two remained in touch, because Rømer was probably the first person to calibrate a thermometer, in 1701, and he used Newton's proposed fiducial points. Rømer never wrote an account of his method, but one of his visitors did: a German instrument maker from Amsterdam called Daniel Gabriel Fahrenheit (1686–1736).

Back home in Amsterdam, Fahrenheit began to experiment. His first thermometer contained alcohol, but he made a significant

CALIBRATING A THERMOMETER

Early thermometers were of the glass-liquid type, using mercury or colored alcohol inside a very narrow tube. This type of thermometer is still widely used, although the use of mercury is being phased out. Today there are several alternatives. Some thermometers use strips of two dissimilar metals fastened together. The metals respond to temperature changes by expanding or contracting, but at different rates, causing the bimetal strip to bend. The movement is transmitted to a needle on a dial. Two half loops of dissimilar metals may also be joined at their ends to form a circle called a *thermocouple*. Changes in temperature cause an electric current to flow through the metals. Ceramic semiconductors also respond electrically to changes in temperature.

All thermometers must be calibrated before they can be used, and from time to time the calibration must be checked, because thermometers can become inaccurate. Calibration involves marking the reading at two temperatures, called fiducial points. The most commonly used fiducial points are the freezing and boiling temperatures of water.

Freezing and boiling temperatures vary with atmospheric pressure, and the water must be pure, because aqueous solutions freeze and boil at slightly different temperatures from pure (distilled) water. Atmospheric pressure varies with altitude and also with changing weather conditions. At change from earlier types, in which the bulb was filled with air so temperature changes made the air expand or contract, pushing the liquid along the tube. Fahrenheit filled the bulb with liquid, so his thermometer measured the expansion and contraction of the liquid, not the air. The thermometer was still not very satisfactory, however, because pure alcohol boils at 172.94°F (78.3°C), so an alcohol thermometer cannot measure very high temperatures. He tried a mixture of alcohol and water, which was better, but introduced a new problem, because the mixture did not expand at a constant rate as the temperature rose, which made calibration difficult. Finally, in about 1714, Fahrenheit used mercury, and devised the final version of the temperature scale that bears his name and used it to calibrate a thermometer of his own making.

That was only part of his achievement, however. Fahrenheit made thermometers that all gave the same reading. Until then, no two thermometers could be relied upon to show the same amount of temperature change.

Fahrenheit described Rømer's method. Rømer, he said, used a thermometer filled with red wine. He inserted it into freezing water

sea level, for example, where the standard air pressure is 29.92 inches of mercury (101.325 kPa), distilled water boils at 212°F (100°C), but at 5,000 feet (1,525 m) altitude, where the average pressure is 25 inches of mercury (85 kPa), water boils at 202°F (94.4°C). At standard sea-level air pressure, seawater containing 35‰ (= 3.5 percent) of salt boils at 213°F (100.56°C) and freezes at 28.56°F (-1.91°C).

Standard sea-level atmospheric pressure is an average value, and the pressure on any particular day is likely to be different. Before calibration can begin, it is essential to know the precise atmospheric pressure so that a correction can be made to the fiducial points.

Calibrating a glass thermometer begins by filling a cylinder half with crushed ice made from distilled water and half with distilled water almost at freezing temperature. The cylinder is then placed inside a larger container lined with solid ice or ice cubes. The ice and water are allowed to stand for about five minutes, to stabilize, and the thermometer is inserted into the cylinder, taking care that no part of it touches the side of the cylinder. After about two minutes the position of the liquid in the thermometer is marked on the tube using an indelible marker. After correction for air pressure, that sets the lower fiducial point. The upper fiducial point is determined by inserting the thermometer into a quantity of vigorously boiling distilled water. The distance between the two fiducial points is then divided into a convenient number of equal units.

and marked the level of wine in the tube. Then he placed the thermometer in tepid water, which Fahrenheit said was at blood heat, and marked the position of the wine. Historians differ about what he did next. Some say he added a third value equal to half the distance between the two fiducial points, but located below the lower point, and marked this as zero. Others say he mixed water, ice, and ammonia to produce a low temperature that he called zero. Yet others say he used melting snow and called that temperature 7½. Whatever his method, in the resulting Rømer scale water freezes at 7½°Rø, boils at 60°Rø, and average body temperature is 22½°Rø.

Fahrenheit aimed to avoid negative values and unnecessary fractions, so he used a mixture of ice and salt to determine his lower fiducial point, calling it zero. He measured body temperature for his upper fiducial point, and divided the distance between the two fiducial points into 90 degrees. On this scale water freezes at 30° and average body temperature is 90°.

Later, using a mercury thermometer capable of registering higher temperatures, Fahrenheit altered his scale. This time he took the boiling temperature of pure water for his upper fiducial point, marked the freezing temperature of pure water, but still retained his lower fiducial point. He divided the distance between the boiling and freezing temperatures into 180 degrees. On his new scale water freezes at 32°, boils at 212°, and average body temperature is 96° (later adjusted to 98.6°). That is the Fahrenheit scale that has remained in use ever since. It is still used in Britain and the United States, although Britain increasingly uses the Celsius scale, bringing the country into line with the other members of the European Union and most other countries in the world (see "Anders Celsius and His Temperature Scale" on pages 63–66).

One other temperature scale was used for a time. Apparently unaware of Fahrenheit's work, in 1730 the French naturalist and physicist René-Antoine Ferchault de Réaumur (1683–1757) devised a scale on which water freezes at 0°R and boils at 80°R.

Daniel Gabriel Fahrenheit (or Gabriel Daniel, there is some doubt about the order of his given names) was born on May 14, 1686, in Danzig, now called Gdansk, a city at the mouth of the Vistula River, on the coast of the Baltic Sea. Today Gdansk is deep inside Poland and culturally it is Polish, but at various times in the past shifting frontiers have meant that it lay inside Prussia or, more recently, Germany. When

Daniel was born it was in the Polish-Lithuanian Commonwealth. The Fahrenheits were merchants, originally from Germany, so culturally they were German and German was the language they spoke. Daniel was the eldest of five children (two boys and three girls) of Daniel and Concordia Fahrenheit who survived childhood. Daniel began his education in Danzig, but both of his parents died on the same day, August 14, 1701, from eating poisonous mushrooms. Following their death the city council assumed responsibility for the children. They placed four of them in foster homes and apprenticed Daniel to a merchant, who took him to Amsterdam. Daniel completed four years of his apprenticeship and learned bookkeeping, but then he became interested in physics and determined to become a maker of scientific instruments. He learned glassblowing, but then the authorities in Danzig heard that he had broken his apprenticeship contract and tried to have him arrested. Daniel went on the run, but when he reached the age of 24, in 1710, he was legally an adult and no longer at risk of arrest.

In about 1707 Daniel left Amsterdam to escape the authorities. Once traveling he visited scientists and instrument makers in Denmark, Sweden, Germany, and Poland. It was while he was in Denmark that he met Ole Rømer.

Fahrenheit returned to Amsterdam in 1717 and established his own business making thermometers, barometers, and other instruments, and lecturing on chemistry. Daniel Fahrenheit remained in the Netherlands for the rest of his life and died in The Hague on September 16, 1736.

ANDERS CELSIUS AND HIS TEMPERATURE SCALE

Daniel Fahrenheit's introduction of a reliable temperature scale did not prevent one of the most distinguished scientists of the day from seeking improvements. In 1742 Anders Celsius, professor of astronomy at the University of Uppsala, Sweden, published a paper, "Observations on Two Persistent Degrees on a Thermometer" in the *Annals of the Royal Swedish Academy of Science.*

Nowadays astronomers study other planets, stars, galaxies, and interplanetary and intergalactic space, but in the 18th century more was expected of them. Navigation at sea depended on accurate and updated information about the night sky. Governments used measurements of the Sun, Moon, and stars to survey territory and define international frontiers. Astronomers supplied the necessary information and devised the techniques for these tasks. Meteorology, concerned as it is with the sky and the changing behavior of the Sun, also interested astronomers. So Celsius was not straying beyond the boundaries of his scientific discipline when he devised his improved temperature scale.

Celsius pointed out in his paper that although water freezes and ice melts always at the same temperature, identifying the precise moment—and, therefore, temperature—at which water begins to freeze is quite difficult. It is much easier to determine the moment at which snow begins to melt. Consequently, he proposed that the lower fiducial point on the temperature scale should be the *melting* point of snow. He said he had checked this temperature very carefully, using a thermometer made by René-Antoine Ferchauld de Réaumur (1683–1757) to measure the temperature of melting snow many times over the course of two winters and under different atmospheric pressures. Nothing if not thorough, he asserted that snow melts at the same temperature at 60°N in Uppsala, 49°N in Paris, 66°N in Torneå, Sweden, and in front of the fire in his own sitting room. He was confident, therefore, in proposing this as his lower fiducial point.

Celsius agreed that the upper fiducial point should be the boiling temperature of water, but this was more difficult to determine. Once water is boiling its temperature ceases to rise, but Celsius suspected that the vigor with which it boiled might affect the temperature, and he noted that when he removed a thermometer from boiling water the mercury rose for a short time before beginning to fall. He attributed this to the glass of the thermometer contracting on contact with the air before the mercury began to do so, consequently shrinking the diameter of the tube and forcing the mercury to rise. He also noted, as had Daniel Fahrenheit, that the temperature of boiling varies according to the atmospheric pressure. Celsius found a way to correct for this by relating the height of the mercury in the thermometer to the height of the mercury in a nearby barometer.

With these two fiducial points identified, Celsius proposed a method for calibrating thermometers. First he determined the lower fiducial point by immersing the thermometer bulb in melting snow. Then he placed the thermometer bulb in boiling water when the atmospheric pressure was precisely 29.75 inches of mercury (100.658 kPa). Finally he divided the distance between the two fiducial points into 100 equal degrees, so that 0° corresponded to the boiling point of water and 100° to the melting point of snow.

In 1750, after Celsius had died, his temperature scale was reversed to the one used today, in which snow melts at 0° and water boils at 100°. Just who altered it is unclear. It might have been Martin Strömer (1707–70), a pupil of Celsius who succeeded him as professor of astronomy. Carolus Linnaeus (Carl von Linné) (1707–78), the professor of botany at Uppsala, also claimed not only to have reversed Celsius's scale, but to have invented the scale in the first place. Linnaeus ordered a mercury thermometer from Daniel Ekström, a famous instrument maker in Stockholm (and Martin Strömer's brother-in-law), possibly in 1743, but the instrument was broken on its way to him. However, a Linnaeus thermometer was installed in the Uppsala University Botanical Garden in November 1745, with the reversed scale. In a description of the orangery in the Botanical Garden that he wrote for the journal *Hortus Upsaliensis* in December 1745, Linnaeus stated that: "Our thermometer shows 0 (zero) at the point where water freezes and 100 degrees at the boiling-point of water." It may well have been Ekström who made the change.

Anders Celsius was a Swedish astronomer who, from 1730, was professor of astronomy at the University of Uppsala. He also invented the temperature scale that bears his name. *(Science Photo Library)*

Anders Celsius was born at Uppsala on November 27, 1701, into a family of scientists. His father, Nils Celsius (1658–1724), was professor of astronomy at the University of Uppsala, his paternal grandfather, Magnus Celsius (1621–79), had been professor of mathematics, and his maternal grandfather, Anders Spole (1630–99), was professor of astronomy prior to Nils. Several of Anders's uncles were also scientists. Anders was educated in Uppsala, showing a great talent for mathematics, and in 1730 he succeeded his father as professor of astronomy. The illustration shows him at about this time.

In 1732 Anders embarked on a grand tour of Europe, visiting many of the leading astronomers of the day and visiting the most important observatories. Sweden had no observatory of international standing, and Anders determined to persuade the government to fund the building of one. He purchased astronomical instruments during his travels, in anticipation of having an observatory to house them.

While he was in Paris, Celsius met the French astronomer Pierre-Louis Maupertuis (1698–1759), who persuaded him to join the expedition he was planning to Torneå, in Lapland, then in northern Sweden (now on the border between Sweden and Finland). The purpose of the expedition was to measure the length of one degree of longitude near to the North Pole, so this measurement could be compared with another made close to the equator in a region of what was then Peru and is now in Ecuador. The overall enterprise aimed to determine the shape of the Earth. The Lapland expedition set off in 1736, returning in 1737, and its results confirmed Sir Isaac Newton's (1642–1727) belief that the Earth is flattened at the poles.

Celsius was recruited to the expedition mainly as a guide and interpreter, but his inclusion was also recognition of his own scientific ability and that Swedish science was of international quality. It made Celsius famous in his own country and helped him persuade the government to build a national observatory. The Celsius Observatory in Uppsala opened in 1741 with Celsius as its first director. The Observatory building is still preserved.

With his assistant Olof Hjorter (1696–1750), Anders Celsius discovered that aurorae are magnetic phenomena, publishing this work in Nuremberg in 1733, based on 316 observations of the aurora borealis he and his colleagues made between 1716 and 1732. He calculated the magnitude of 300 stars. He strongly supported efforts to introduce the Gregorian calendar to Sweden. This had been tried in 1700 by omitting the leap days between 1700 and 1740, but 1704 and 1708 were incorrectly declared leap years, and in 1712 Sweden reverted to the Julian calendar. The Gregorian calendar was finally introduced in 1753, the change being effected by deleting 11 days from the calendar.

Anders Celsius died on April 25, 1744, from tuberculosis. He is buried next to Magnus Celsius at Gamla Uppsala (Old Uppsala), a village about 3 miles (5 km) from the center of Uppsala.

Water in the Air

There is no weather on the Moon. Several commercial organizations are planning to build tourist facilities on the Moon, and one day wealthy vacationers may fly to the lunar resort of their choice, where they will be able to gaze upon a landscape utterly different from any scene on Earth. But they will not wake each morning wondering whether the day will be fine or wet, whether the sky will be clear or cloudy. The sky will be black all day every day. There will be no weather because the Moon lacks an atmosphere. When the lunar tourists walk outdoors they will need to take air with them, for the Moon has none.

There can be no weather without air, but it is not really the air that produces the weather, but water. Will the day be fine or wet? That depends on water. Will the sky be clear or cloudy? Clouds are made from water. Will the harvest be a good one? Crops need sunshine (skies with relatively little water), but also rain—in other words water again. In early February 2008 transport and power systems were disrupted all over China, just as people were trying to get home to celebrate the New Year with their families. The winter was intensely cold, but it was not the cold that caused the problems, but the heavy snowfall. Snow is water, of course.

Even the wind requires water. Wind blows to equalize differences in air pressure, but those differences occur because the evaporation and condensation of water absorb or release energy into the air, making it expand or contract (see "Joseph Black, Jean-André Deluc, and

Latent Heat" on pages 109–112). It is the phase changes of water that drive weather systems. Hurricanes, which are the most dramatic and violent of tropical weather systems, derive all of their energy from the condensation of water vapor, and the amount of energy an average hurricane releases is equivalent to about 200 times the electrical generating capacity of the entire world.

This chapter discusses atmospheric water. It describes how scientists first discovered a way to measure the amount of water present in the air and explains what it is that they were measuring. It tells of studies of the way clouds form and the origin of the names we use to identify each type of cloud. Finally it explains what is happening during a thunderstorm and describes Benjamin Franklin's famous—and extremely risky—experiment that demonstrated that lightning is electrical.

GUILLAUME AMONTONS AND THE HYGROMETER

The learned scholars who attended meetings at the Accademia del Cimento in the middle of the 17th century to explore weather phenomena knew that the air contains water vapor and that the amount varies from day to day. The amount of water vapor present in the air is known as the humidity and there are several ways to measure and report it (see sidebar below).

It was Grand Duke Ferdinando II de' Medici (1610–70) who made what was probably the first instrument to measure atmospheric moisture. As the following illustration shows, his instrument consisted of a vessel, a funnel, and a flask, and it was elaborately and beautifully decorated with pictures of people holding scientific instruments. It could not tell the experimenter how much moisture the air was holding, but it could tell whether this was a relatively large or small amount. The central vessel would be filled with ice. Water vapor then condensed onto the outside of the vessel and trickled down into the funnel, and from there into the flask. The amount of water in the flask indicated the relative amount of moisture in the air.

An instrument for measuring the amount of water vapor in the air is called a hygrometer, from the Greek word *hugros* meaning wet, and although Ferdinando's may have been the first, other physicists also experimented with instruments to measure water vapor. Most

HUMIDITY

Humidity is a measure of the amount of water vapor present in the air. It refers only to water vapor, which is a gas, and not to droplets of liquid water, for example in clouds. It is measured using a hygrometer or *psychrometer*. A psychrometer uses two thermometers to measure air temperature. One thermometer has its bulb directly exposed to the air and the bulb of the other thermometer is wrapped in a muslin wick, the end of which is immersed in a reservoir of water, so water is drawn into the fabric. Water evaporating from the muslin around the bulb draws heat from the bulb, so the *wet-bulb thermometer* registers a lower temperature than the *dry-bulb thermometer*. The difference between the two readings, called the wet-bulb depression, can be used to calculate the humidity and also the dew-point temperature, which is the temperature at which water vapor will condense.

Humidity is measured and reported in several different ways, as the mixing ratio, specific humidity, absolute humidity, or relative humidity.

The mixing ratio, also called the mass mixing ratio, is the ratio of the mass of any particular gas present in the air to a unit mass of air without that gas. It is usually expressed as grams of the gas per kilogram of air without the gas. When used to express humidity it is given as grams of H_2O per kg of air (gH_2O/kg).

Specific humidity is the ratio of the mass of water vapor present in the air to a unit mass of air including the water vapor. Despite comparing the mass of water vapor with moist rather than dry air, in practice there is little difference between the mixing ratio and specific humidity. This is because even in very moist air, water vapor seldom accounts for more than a very small proportion of the total mass, so including or omitting it makes little difference.

Absolute humidity is the mass of water vapor present in a given volume of air, usually expressed as grams per cubic meter (g/m^3). This takes no account of the fact that changes in temperature and pressure alter the volume of air, and hence the absolute humidity, without adding or removing water vapor. This means that absolute humidity is a somewhat unreliable measure, and it is little used.

Relative humidity (RH) is the measure most often used and the one that is always used in radio and TV weather reports and forecasts. It is the amount of water vapor present in the air expressed as the percentage of the amount of water vapor that would be needed to saturate the air at the prevailing temperature. Since the amount of water vapor air can hold varies with the temperature (the lower the temperature the less water vapor air can hold), RH varies according to the temperature. This makes RH a measure of limited value to meteorologists, who need to know the amount of water vapor rather than how close the air is to saturation. It is a simple measure, however, easy to determine and express, and useful in weather forecasting because it indicates the likelihood of cloud formation and precipitation. RH is reported either as a percentage (e.g., "The humidity today is 78 percent.") or as a decimal fraction (e.g., "The humidity today is 0.78.").

Fernandino II's hygrometer. The central vessel was filled with ice. Water vapor condensed on the outside of the vessel and trickled down into the funnel and from there into the flask, where it was collected. The amount of water in the flask indicated whether there was a large or small amount of moisture in the air.

of these exploited the fact that many natural substances expand and contract as they absorb and lose atmospheric moisture. Seaweed and pine cones are familiar examples, and various experimenters also used salt, hemp, catgut, oat fibers, or wood. As these substances absorb moisture their weight increases, and they become lighter as the moisture evaporates into dry air. Changes in weight indicate the amount of moisture in the air—but not very accurately.

Francesco Lana de Terzi (1631–87), an Italian Jesuit at the Academy of Nuremberg, invented a hygrometer that contained a long piece of plant fiber concealed inside a cylinder. The lower end of the fiber was fastened to the bottom of the cylinder, and as the fiber's length changed with the changing humidity, it moved an arrow pivoted at its center and held by a figure at the top of the cylinder. The diagram below shows how it worked. Francesco Lana de Terzi also designed a lighter-than-air flying machine that he called a flying ship.

In 1687 the French physicist Guillaume Amontons (1663–1705) made a much more ingenious hygrometer. Amontons also based his design on the capacity of certain materials to absorb and lose atmospheric moisture, but he used this in an entirely novel way. At the center of the Amontons hygrometer there was a hollow sphere made of leather—he also used beechwood and horn. The sphere was filled with mercury and a glass tube rose vertically from the top. The size of the sphere changed as it absorbed and lost moisture. As it shrank, it squeezed mercury into the tube and as it expanded mercury fell from the tube. The level of mercury in the tube indicated the humidity of the air, although with no units with which to quantify the measurement.

Guillaume Amontons was one of the most talented inventors of his generation. He was born in Paris on August 31, 1663, the son of a lawyer who came originally from Normandy. Guillaume attended the Latin school in Paris, where he studied the physical sciences, celestial mechanics, mathematics, drawing, surveying, and architecture. So far as is known he did not attend a university. After leaving school he was employed by the government on a number of public works. This gave him valuable experience of mechanics, and after a few years as a government employee he devoted himself to designing and improving scientific instruments.

While still at school Amontons fell seriously ill with a complaint that left him profoundly deaf. This made communication with other people extremely difficult, but far from regarding it as a handicap,

Francesco Lana de Terzi's hygrometer. The vegetable fiber is secured to the base of the cylinder. Its length changes as it absorbs water from moist air and loses water by evaporation into dry air. As this happens it moves the arrow, which is free to rotate.

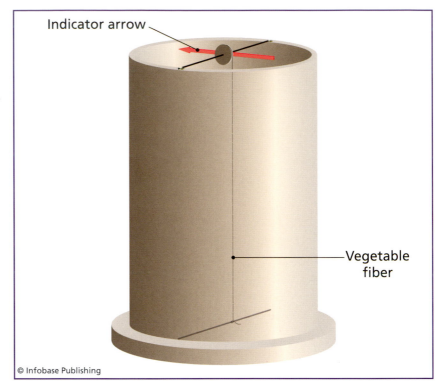

Indicator arrow

Vegetable fiber

© Infobase Publishing

Amontons considered it a blessing, because it allowed him to concentrate on his scientific work without fear of distraction.

Guillaume Amontons was only 24 years old when he presented his new hygrometer to the Académie des Sciences, where it was warmly received. The following year he developed an optical telegraph system, in which a person on one hilltop signaled with a bright light to a recipient with a telescope on another hilltop. He demonstrated his system to the king some time between 1688 and 1695, but it was never adopted. He invented a fire engine in which a fire heated water held inside the hollow rim of a wheel, causing the water to expand on one side of the wheel, making the wheel rotate. This led him to study friction and its effect on moving machinery. In 1695 Amontons invented a new type of *clepsydra*—a clock operated by moving water. He thought this would be useful at sea, but it was insufficiently accurate for calculating longitude. Amontons improved on the air thermometer, using mercury and adjusting the height of the mercury column until the air occupied a fixed volume.

After that, changes in the temperature of the air altered the pressure the air exerted on the mercury, so the temperature of the air was all that the instrument registered. His experiments with the air thermometer led him to speculate that there might be a temperature so low that the air pressure reached zero. This was the first suggestion of the concept of absolute zero temperature. Amontons invented several barometers, including one without a mercury reservoir that could be used on ships, and he used a barometer as an altimeter.

In 1690, on the recommendation of the astronomer Jean Le Fèvre (1652–1796) Guillaume Amontons was appointed a member of the Académie des Sciences. He published a number of papers and, in 1695, *Remarques et expériences physiques sur la construction d'une nouvelle clepsydre, sur les baromètres, thermomètres, et hygromètres* (Observations and physical experiences on the construction of a new clepsydra, on barometers, thermometers, and hygrometers), his only book, in which he described his inventions.

Early in October 1705 Amontons suffered an acute attack of peritonitis. He died, in Paris, on October 11.

JOHN FREDERIC DANIELL AND THE DEW-POINT HYGROMETER

There is one air temperature that is familiar to everyone: the dew point. This is the temperature at which water vapor condenses and liquid water begins to evaporate. Dew forms naturally on clear nights when the air is fairly still. During the day, surfaces exposed to sunlight absorb the Sun's radiation and their temperature rises. At the same time, they emit their own radiation, losing the warmth they have absorbed. By day surfaces absorb warmth faster than they lose it, but at night when there is no solar warmth to absorb, they continue to lose heat by radiating it upward. Their temperature begins to fall in late afternoon and continues falling until about an hour before dawn. On cloudy nights the clouds absorb this outgoing radiation, their droplets and water vapor re-radiating it in all directions, some of it downward. This warms the air below the cloud and it is why cloudy nights are generally warmer than clear nights.

As surfaces radiate away their absorbed heat, their temperature falls and the thin layer of air in direct contact with the surfaces also becomes chilled (for an explanation why, see "Joseph Black, Jean-André Deluc, and Latent Heat" on pages 109–112). This effect is strongest on clear nights, when the temperature in this *boundary layer* of air may fall low enough for the water vapor it contains to saturate it. The relative humidity in the boundary layer reaches 100 percent, and water vapor condenses onto the surfaces, forming dew.

If the air is quite still, condensation will take place only in the boundary layer, and the dew will be light. If the air is moving gently, the turbulence this produces will draw down moist air from a higher level and its water vapor will also condense, adding to the amount of dew. Clear nights with a very light breeze often produce heavy dew. A stronger breeze will bring in warmer air, however, preventing dew from forming at all.

Dew will also form by contact cooling if warm, moist air moves across a much cooler surface. Contact with the cool surface chills the air and water vapor will condense if the temperature in this layer of air falls below the dew point. This type of cooling will also produce *advection fog.*

The *dew-point temperature* varies according to the amount of water vapor in the air. Consequently, it is possible to calculate the amount of water vapor in the air—the mixing ratio or specific humidity (see sidebar "Humidity" on page 69)—if the dew-point temperature is known. This calculation is based on the difference between the air temperature outside the boundary layer and the dew-point temperature, a difference known as the *dew-point depression.*

In 1820 the English chemist, physicist, and inventor John Frederic Daniell (1790–1845) invented an instrument that measured the dew-point temperature and air temperature simultaneously, allowing the dew-point depression to be calculated by simple subtraction. His dew-point hygrometer remained in use for many years; an electronic instrument has now taken its place.

The following illustration shows how Daniell's dew-point hygrometer worked. It consisted of a closed glass tube with a bulb at each end, and one bulb held higher than the other. The lower bulb contained liquid ether and a thermometer in this arm of the tube had its bulb inside the ether bulb and partly immersed in the ether. The other bulb was wrapped in muslin. An ordinary thermometer registering air temperature was mounted on the center of the stand. Liquid ether was poured onto the muslin, saturating it. The ether evaporated rapidly, lowering the temperature inside the bulb. As the temperature fell, ether vapor inside the tube started to condense in the bulb. This reduced the *vapor pressure* in the lower bulb, causing its ether to start evaporating. That evaporation chilled the bulb and when its temperature fell below the dew-point temperature, atmospheric water vapor began to condense on the outside of the bulb. The thermometer inside the bulb registered the temperature at which this happened, and the central thermometer registered the ambient air temperature at the same moment.

John Frederic Daniell was one of the most eminent scientists of his day and the inventor of an electric battery known as the Daniell cell, a barometer, and a *pyrometer*, as well as the dew-point hygrometer. A pyrometer measures the temperature of an object from the radiation it emits. Daniell was born in London on March 12, 1790. His father was a lawyer. John was educated privately, mainly studying Latin and Greek. It is uncertain whether he earned a degree from the University of Oxford or was awarded an honorary degree.

His education gave him a good grounding in technology, and John went to work for a relative who owned a sugar refinery and resin factory, where he was able to improve on the technology being used. In his spare time John attended chemistry lectures at the Royal Institution given by Professor William Thomas Brande (1788–1866). Daniell met Brande and the two became friends. Between them they revived the fortunes of the Royal Institution, which were then at a low ebb.

In 1813, at the age of 23, Daniell was appointed professor of physics at the University of Edinburgh. In 1823 he was elected a fellow of the Royal Society. He left Edinburgh in 1831 to take up a position as the first professor of chemistry and meteorology at King's College, London, a position he held until his death.

While he was at Edinburgh, Daniell combined his teaching of physics with work as a chemist and for a time in 1817 he managed the Continental Gas Company, where he developed a new process for manufacturing gas by dissolving resin in turpentine and distilling gas from the solution.

John Daniell was also keenly interested in meteorology. It was in 1820 that he invented his hygrometer and in 1823 he published *Meteorological Essays,* a book in which he described his meteorological research. He investigated the influence of solar radiation on the

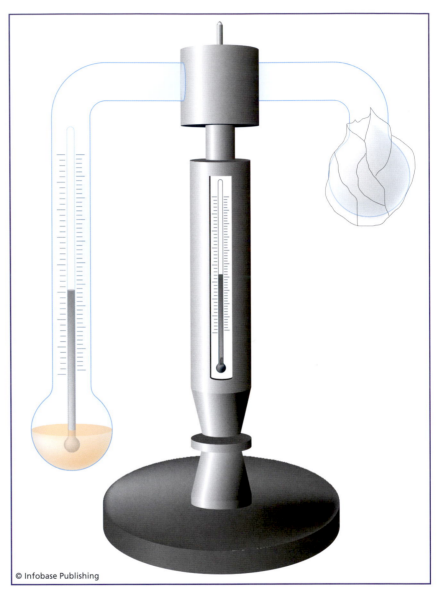

© Infobase Publishing

John Daniell's dew-point hygrometer. The bulb on the left contains ether and has a thermometer bulb partly immersed in it. The bulb on the right is ordinarily wrapped in muslin. Liquid ether is sprinkled over this bulb. The ether evaporates, lowering the air temperature inside the bulb and ether vapor starts to condense. The lowering of vapor pressure makes ether evaporate from the left-hand bulb, lowering its temperature. Water vapor condenses onto the left-hand bulb when its temperature falls to the dew-point.

climate and studied the circulation of the atmosphere, and he was the first person to recognize the importance of maintaining a moist atmosphere inside greenhouses. This revolutionized greenhouse horticulture, and in 1824 Daniell was awarded the silver medal of the Horticultural Society (now the Royal Horticultural Society). In the same year he published *Essay on Artificial Climate Considered in Its Applications to Horticulture.*

His interest in chemistry and physics had not diminished, and, in the 1830s, through his friendship with Michael Faraday (1791–1867), Daniell became increasingly interested in electrochemistry. This led in 1836 to his invention of the Daniell cell, which was the first reliable source of direct-current electricity. John Daniell published *Introduction to the Study of Chemical Philosophy* in 1839 and in 1841 he became a founding member and vice president of the Chemical Society of London.

Daniell was very active in the affairs of the Royal Society. In 1830 he installed one of his own barometers in its entrance hall. This instrument used water to measure atmospheric pressure, and Daniell took many readings from it. He was foreign secretary of the Society from 1839. On March 15, 1845, while attending a Royal Society meeting, John Daniell suffered a heart attack and died.

HORACE-BÉNÉDICT DE SAUSSURE, THE HAIR HYGROMETER, AND THE WEATHER HOUSE

Many natural fibers absorb atmospheric moisture when the air is humid and lose it in drier air, and one of those fibers is human hair. As the hair absorbs and loses moisture, its length changes and the extent of this change is always the same for a particular hair. Between a relative humidity of 0 percent and one of 100 percent the length of a human hair changes by 2–2.5 percent, becoming longer as the humidity increases and shorter as it decreases. Household hygrometers, which have a needle and dial to display relative humidity, exploit this phenomenon. They are hair hygrometers.

The first person to make use of this change in hair length—due to the physical structure of hairs—was the Swiss physicist Horace-Bénédict de Saussure (1740–99). In 1783 he made the first hair hygrometer, and he published a description of it in his book *Essais sur l'Hygrométrie, 1er Essai, Description d'un nouvel Hygromètre comparable* (Essays on hygrometry, 1st essay, description of a new comparable hygrometer), published in Geneva in 1788. His instrument contained a human hair secured inside a frame at one end and wound around a worm screw at the other end. The worm screw was attached to a needle that moved around a dial. The hair was held under tension by a cord wound around the axle of the needle and the amount of tension could be adjusted. As its length changed in response to changing humidity, the hair turned the worm screw, which made the needle move. The dial was divided into 360 divisions, each subdivided into 10 smaller divisions, so de Saussure had, for the first time, assigned a numerical value to humidity. He measured it in degrees and fractions of a degree.

De Saussure calibrated his hygrometer by determining the two extremes of 100 percent and 0 percent relative humidity. He produced 100 percent humidity by placing the hygrometer beneath a bell jar containing a receptacle of water and waiting while evaporation saturated the air. To dry the air he used the same bell jar, but this time with a quantity of a hygroscopic (water-absorbing) compound such as sodium chloride (common salt) or sodium hydroxide (lye).

Scientific instruments are often elaborately decorated and they can be very beautiful, but the hair hygrometer has gone several steps further and become a household ornament that is almost a toy. The weather house is a hair hygrometer, but instead of moving a needle against a dial, changes in the length of the hair move two figures, so one advances while the other retires through two doors in the front of a house. Depending on which of the two figures stands outside the house, the weather will be wet or fine. The house itself is often meant to resemble a Swiss chalet, perhaps in memory of de Saussure. The illustration that follows shows a typical example.

Horace-Bénédict de Saussure was born at Conches, near Geneva, on February 17, 1740. In 1746 he enrolled at the public school in Geneva and advanced to the Geneva Academy in 1754. He graduated from the Academy in 1759, after delivering a dissertation on the physics of fire.

In 1760 de Saussure made his first visit to Chamonix, across the border in France and at the foot of Europe's highest mountain, Mont Blanc, towering to 15,771 feet (4,810 m). No one had ever climbed it, and de Saussure, fascinated by the mountain, toured the nearby

A weather house is meant to predict the weather. The house has two doors, with a man behind one and a woman behind the other. Depending on which of them emerges from the house, the weather will be wet or fine. A thermometer between the doors completes the device. In fact, a weather house is a hair hygrometer. *(Kurt Kormann/Corbis)*

parishes offering a substantial reward to the first person to reach the peak. Jacques Balmat and Gabriel Paccard were the first people to climb all the way to the summit, on August 8, 1786. De Saussure climbed the mountain the following year, reaching the summit at 11 A.M. on August 3, 1787, accompanied by several guides and his personal valet. Over the succeeding years he returned to the mountain many times and climbed other alpine mountains. His interest was not primarily in mountaineering, however, but in geology, meteorology, and alpine plants. His three-volume work *Voyages dans les Alpes* (Travels in the Alps) appeared between 1779 and 1796 and may have been the first books to contain the word *geology*.

The air is always cooler near the peak of a tall mountain than it is in the valley below and de Saussure determined to find out why. He needed a device for his experiment and in 1767 made what may have been the very first solar collector. This was a sealed box, heavily insulated, with a glass top, containing a thermometer. He took his box to the summit of Mont Cramont, more than 8,000 feet (2,440 m) above sea level, where the air temperature was 43°F (6°C). The thermometer inside his box registered 190°F (88°C), however. De Saussure then repeated his observation on the Plains of Cournier, about 4,850 feet (1,480 m) lower down. There the air temperature was 77°F (25°C), but the temperature inside the box remained close to 190°F (88°C). De Saussure concluded that the Sun shines just as strongly on the plains as it does high in the mountains, but that the more transparent air at high elevations is unable to trap and hold so much of the heat.

In 1761 de Saussure applied unsuccessfully for the position of professor of mathematics at the University of Geneva. The following year he applied again, this time successfully, for the professorship of philosophy. In 1772 he was elected a fellow of the Royal Society.

De Saussure married Albertine Boissier in May 1765. In February 1768, accompanied by his wife and sister-in-law, he visited Paris, the Netherlands, and England, returning to Geneva in 1769. In 1771 and 1772 he toured Italy, on the second journey taking his wife and their six-year-old daughter.

These were troubled times politically, as Switzerland felt the effects of the events leading up to the French Revolution of 1789. De Saussure became involved in plans for democratic reform. He was briefly arrested in 1782, and the same year he was besieged for several days in his own home, suspected of concealing weapons and harboring troublemakers. Geneva had its own revolution in 1792, and in 1794 the Terror that followed the Revolution in France spread to the city.

By then de Saussure had left. He resigned his position at the Geneva Academy in 1787 and moved to the south of France, where he could live close to sea level and take measurements of the atmosphere that he could compare with those he had made in the Alps. His health had begun to deteriorate in 1772, and by 1794 he was a sick man. He was also poor, and his financial difficulties compelled him to return to the family home at Conches. News of his straitened circumstances spread, and he received offers of assistance from abroad. Thomas Jefferson considered offering de Saussure a position at the newly opened University of Virginia, but it was too late. Horace-Bénédict de Saussure died at Conches on January 22, 1799.

MEASURING RAINFALL

The amount of annual rainfall is closely linked to the size of a harvest, and around 400 B.C.E. taxes in parts of India were levied on the amount of grain produced. The tax collectors did not need to trust farmers to declare the size of their harvest truthfully, because they had a way to calculate how much should have been harvested. They measured and recorded the amount of rainfall, computed the yield, and based their tax levels on their computations.

They were not the first people to measure rainfall. Rainfall was being recorded a century earlier in Greece, although Aristotle did not mention how rainfall was measured in his book *Meteorologica* (see "Aristotle and the Beginning of Meteorology" on pages 2–6).

The first instrument to measure rainfall in a standardized way, so records from one instrument could be readily compared with those from another, was probably made in Korea in 1441, during the reign of King Sejong (1397–1450). It took the form of a container of standard size, about 5.5 inches (14 cm) in diameter and 12 inches (30 cm) deep, mounted on top of a pillar. The purpose was to help farming communities predict their harvests, and, therefore, to inform government officials about how much tax each community should pay. Every village received one of these rain gauges.

Two types of rain gauges are widely used in modern weather stations. The standard gauge is extremely simple, consisting of a graduated collecting cylinder with a funnel on top. The tipping-buckets gauge is a little more complicated. The following illustration shows both.

No one knows who invented the standard gauge, but by the middle of the 19th century meteorologists were working to standardize its dimensions. Standardization was clearly necessary, because the gauge measures the depth of rainfall, in inches or millimetres. Unless all gauges are of similar diameter, it will be impossible to compare their readings. The international standard for a rain gauge of this type was first established in the United States in 1887 by Cleveland Abbe (1838–1916), then the scientific assistant at the weather service established in 1870 as part of the Army Signal Service. Abbe specified the standard in a paper on "Meteorological Apparatus and Methods" that was published in the 1887 Annual Report of the Chief Signal Officer. With the dimensions translated into scientific units, Abbe's standard became the international standard.

The two types of rain gauge that are widely used at weather stations. The standard gauge (left) has to be visited at regular intervals to be read and emptied. The tipping-bucket gauge (right) allows the rainfall to be measured automatically.

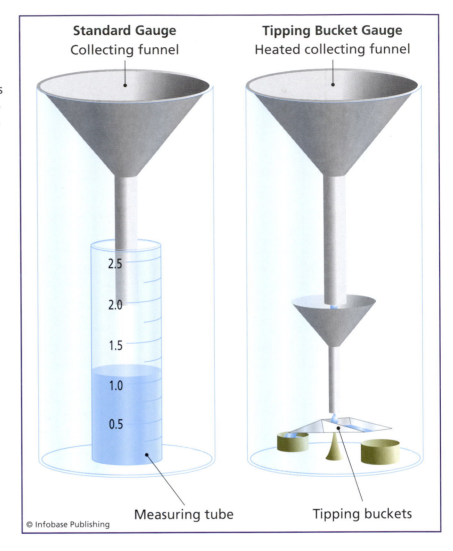

Standard Gauge
Collecting funnel

Tipping Bucket Gauge
Heated collecting funnel

2.5

2.0

1.5

1.0

0.5

Measuring tube

Tipping buckets

© Infobase Publishing

    The standard gauge is a cylinder 7.9 inches (20 cm) in diameter that is mounted vertically with its top 39.4 inches (1 meter) above ground level. A funnel at the top of the cylinder directs rainwater into a measuring tube 2.49 inches (6.32 cm) in diameter. The cross-sectional area of the measuring tube is one-10th the area of the mouth of the funnel. Consequently, the height of water collected in the measuring tube represents 10 times the depth of rain that fell on the ground. This makes the gauge easy to read: 5 inches (or millimeters) of water in the gauge is equivalent to 0.5 inches (or millimeters) of rain on the ground.

The tipping-buckets gauge also consists of a cylinder with a funnel at the top. In this case the funnel feeds rainwater into a second funnel lower down, and the second funnel directs the water into one of two containers called buckets. The buckets are pivoted like a seesaw so that when the one receiving water is full its weight causes it to tip downward and empty its contents, at the same time positioning the second bucket to continue receiving rainwater. Each time the buckets tip the fact is recorded, originally by a system of levers operating a pen, and nowadays by completing an electrical circuit that sends a signal to a computer. Each bucket in a modern tipping-buckets gauge holds the equivalent of 0.007 inch (0.2 mm) of rain and if the gauge is not linked to a computer, rainfall is recorded on a chart fixed to a rotating drum that takes 24 hours to complete one rotation.

Sir Christopher Wren (1632–1723) invented the tipping-buckets rain gauge in about 1661, working in collaboration with his friend Robert Hooke (see "Robert Hooke and the Wheel Barometer" on pages 52–59).

JOHN AITKEN AND THE FORMATION OF CLOUDS

When atmospheric water vapor condenses, it forms minute droplets that gather together as clouds. The process sounds simple, but it is not. In the first place, why is it that clouds form in particular places and at particular heights, but not at others even when the air is equally moist over a wide area? How can it be that sometimes clouds will form over the sea when the relative humidity of the air reaches about 80 percent, but over land the relative humidity sometimes exceeds 100 percent and still no cloud forms in the supersaturated air?

Dew provides a clue—it forms by the condensation of water vapor onto cold surfaces, and surfaces are essential. The dewdrops are on the surfaces of leaves and other objects, but they do not form tiny clouds hovering above them. In 1917 the Royal Society awarded its Royal Medal to the Scottish scientist who solved the mystery of what happens when water vapor condenses. That scientist was John Aitken (1839–1919).

John Aitken studied dust. His published papers were all about dust, with titles such as "On the number of dust particles in the atmosphere" (1888) and "Dust and meteorological phenomena" (1894).

What Aitken discovered was that air contains very large numbers of extremely small particles and that these particles present the surfaces onto which water vapor condenses. The smallest of the particles are called Aitken nuclei in his honor. An *Aitken nucleus* is less than 0.0016 inch (0.4 μm) across, and most are much smaller. The air contains solid and liquid particles, known collectively as *aerosol,* of many sizes, up to giants more than 0.08 inch (20 μm) across, but it is Aitken nuclei that are most effective at encouraging condensation and forming clouds. The particles involved in cloud formation are known as *cloud condensation nuclei* (CCN), and some are larger than Aitken nuclei, which are the smallest. Air must be supersaturated (with relative humidity of about 101 percent) before clouds will form in the absence of CCN.

Aitken invented a device, called the Aitken nuclei counter, for detecting these tiny particles. It is a chamber containing sodden filter paper to make sure the air inside the chamber is saturated. A sample of air is drawn into the chamber, and a pump makes the air expand rapidly. The rapid expansion chills the air, and its water vapor condenses onto particles in the air. Some of the water droplets fall onto a graduated disk, and a lens makes them clearly visible so they can be counted, on the assumption that each droplet represents one Aitken nucleus. The number of nuclei in a unit volume of air can then be calculated from the number counted on the disk.

As physicists began counting the nuclei in air samples collected in different places, not surprisingly they learned that there are many more particles over land than there are over the open sea, far from land. That is to be expected if the particles are mainly dust and include a large proportion of very small soil particles. But then Aitken noticed something else. He found there were many more particles in the air on sunny days than on cloudy days. This is not because clouds attracted particles, leaving the air between clouds depleted, but because the concentration really was different. He also noticed that on sunny days with a high particle concentration the air was clear and not hazy. This told him that the particles were the size of molecules. He concluded that these particles are the products of chemical reactions taking place in the air and driven by the energy of sunlight. He found that air over densely inhabited regions contains the highest concentrations of these particles, describing this as the regions "losing their purity," and he reasoned that condensation coats the particles with water and the droplets merge to form raindrops, thus removing the particles from the air. He believed there are parts of the world "that lose more impurity than they gain."

Meteorologists have not discovered any regions where the air is naturally purifying itself because this process is now known to occur everywhere. It is called *rainout,* and as raindrops fall they tend to sweep up more particles, also removing those in a process called *washout.* As Aitken suspected, rainout and washout are highly efficient mechanisms for cleaning the air. Because of them, pollutant particles such as smoke remain airborne for no more than a few hours. Aitken had also drawn attention to atmospheric chemistry, suggesting *photolytic* (light-driven) reactions that are now known to include those responsible for photochemical smog.

John Aitken was born at Falkirk, Stirlingshire, on September 18, 1839, the son of a lawyer, Henry Aitken of Darroch. John attended Falkirk Grammar School and studied marine engineering at the University of Glasgow. After graduating he worked for a time for R. Napier and Sons, a Glasgow shipbuilding firm. At that time the shipbuilding industry was expanding, and by the end of the century Glasgow shipyards would dominate the global market. Aitken returned to Falkirk, however, where he remained for the rest of his life. His health was always poor, making it impossible for him to hold down a permanent job. Instead, he made himself a laboratory and worked at home.

Aitken was elected a fellow of the Royal Society of Edinburgh in 1875 (proposed by Lord Kelvin) and the University of Glasgow awarded him an honorary doctorate in 1889. Also in 1889 he was elected a fellow of the Royal Society. John Aitken died at his home on November 13, 1919.

TOR BERGERON AND HOW RAINDROPS FORM

Cloud droplets are extremely small. Their size depends on the size of the cloud condensation nuclei onto which water vapor condenses— the smaller the nuclei the smaller the droplets. An average droplet is about 0.0004 inch (10 μm) across, and inside a typical cloud there is an average of 28,300 water droplets in each cubic foot of air (100,000 per liter). This is a large number, but the small size of the droplets means that a cloud is mostly empty air.

Water droplets are subject to gravity, and they fall, but they are so small that they fall quite slowly. Average-size cloud droplets fall at about 0.4 inch per second (1 cm/s), or about 120 feet (37 m) in an hour. The largest droplets, which are about 0.002 inch (50 µm) across, fall faster, at up to about 11 inches per second (27 cm/s), or 3,300 feet per hour. As they fall, cloud droplets lose water by evaporation. When they enter the dry air beneath the cloud or are carried by air currents into the dry air above the cloud, they evaporate completely. That is why clouds have clearly defined edges when seen from a distance.

John Aitken had discovered how water vapor condenses to form clouds (see "John Aitken and the Formation of Clouds" on pages 83–85), but not how cloud droplets can grow large enough to reach the ground before they evaporate. How do clouds produce rain?

The speed with which a water droplet falls—its terminal velocity—is proportioned to 8,000 times its radius. This means that a droplet 0.04 inch (1 mm) in diameter will fall at about 157 inches per second (4 m/s). If the base of the cloud is at a height of 200 feet (60 m), the droplet will take 15 seconds to reach the ground from the time it leaves the cloud. A droplet half that size would take 30 seconds. During that time it would evaporate. The rate of evaporation imposes a limit of about 0.04 inch (1 mm) on the size of raindrops and 0.02 inch (0.5 mm) for droplets of drizzle, which fall from a much lower cloud base. The puzzle is how droplets 0.0004 inch (10 µm) across grow into raindrops 0.04 inch (1 mm) across. One small raindrop is the size of 1 million cloud droplets.

In 1933 the Swedish meteorologist Tor Bergeron (1891–1977) proposed one way in which raindrops can form. The mechanism he suggested was confirmed experimentally in 1938, using large cloud chambers, by the German meteorologist Walter Findeisen (1909–45) and is usually known as the Bergeron-Findeisen mechanism. It had also been suggested as early as 1911 by the German meteorologist Alfred Wegener (1880–1930), so it is sometimes called the Wegener-Bergeron-Findeisen mechanism.

The story began in 1922, when Bergeron spent several weeks at a Norwegian health resort north of Oslo. The resort was often shrouded in fog, and Bergeron noticed that when the temperature was well below freezing there was no fog on the roads through the forest, but the fog extended all the way to the ground when the temperature was above freezing. When the temperature was below freezing the forest trees were covered in frost, and this gave Bergeron a clue to what was happening. He realized that at temperatures below freezing the *saturation vapor pressure* is higher over water than it is over ice. Vapor pressure is the proportion of atmospheric pressure exerted by water vapor, and the saturation vapor pressure is the vapor pressure at which water vapor saturates the thin layer of air immediately adjacent to a surface. Bergeron developed his ideas in his doctoral thesis, which he presented in 1928, and he expanded them in 1933.

The mechanism Bergeron proposed takes place in *cold clouds*. These are clouds that contain both ice crystals and *supercooled* liquid droplets—droplets that remain liquid below freezing temperature. (Raindrops form by a different mechanism in *warm clouds*, which consist entirely of water droplets.) The saturation vapor pressure is higher over liquid water than it is over ice, and the difference is greatest when the air temperature is between about 5°F (-15°C) and -13°F (-25°C). Under these conditions supercooled droplets lose moisture by evaporation, while ice crystals grow by the direct *deposition* of ice from water vapor. As they grow, the ice crystals collide with one another and join to form aggregations—snowflakes. The snowflakes continue growing until they are heavy enough to start falling. They descend through the cloud, colliding with supercooled droplets that freeze onto them, so the snowflakes continue growing as they fall. From time to time collisions shatter large snowflakes, producing small splinters of ice that are carried aloft by air currents and form nuclei onto which more ice is deposited, continuing the process. Large snowflakes fall from the base of the cloud. If the temperature near the bottom of the cloud and in the air between the cloud base and the surface is above freezing the snowflakes melt and become raindrops. If the temperature between the cloud base and the surface is lower than about 39°F (4°C) the precipitation falls as snow. In middle latitudes, even in the middle of summer, most rain consists of snowflakes that have melted during their descent. The Bergeron-Findeisen mechanism predominates.

Tor Harold Percival Bergeron was born on August 15, 1891, at Godstone, Surrey, south of London. His parents were Swedish, and Tor was brought up in Sweden. He graduated in 1916 from the University of Stockholm, after which he conducted research informally at the Swedish Meteorological Institute, helped by the fact that his mother knew the director. On January 1, 1919, Bergeron was appointed as an extra assistant meteorologist. By that time the institute had been renamed the Swedish Meteorological and Hydrological Institute (SMHI). Bergeron was there for only a few months before moving to join Vilhelm and Jacob Bjerknes and their colleague Halvor Solberg at the weather-forecasting service they were establishing at the Bergen Museum (see "Vilhelm Bjerknes and the Bergen School" on pages 149–154). During the autumn of 1919 Bergeron observed that sometimes a *cold front* would seem to overtake a *warm front.* By 1922 he had called this situation an *occlusion* and established its importance in the life cycle of extratropical *cyclones.*

During the 1920s Bergeron spent time studying at the University of Leipzig, Germany. Out of his work there he wrote the thesis for which in 1928 he was awarded his doctorate from the University of Oslo. After receiving his doctorate Bergeron lectured on the methods of the Bergen School in Malta and the Soviet Union.

Bergeron applied in 1935 for the position of professor of meteorology at the University of Uppsala. His attempt failed, partly due to resentment of his close association with the Norwegian meteorologists at Bergen and their methods, which were unfashionable. He returned to Stockholm in 1936 to work at the SMHI, eventually becoming its senior scientist. In 1947 Bergeron obtained the post he had applied for years ago, becoming professor of meteorology and head of the department of synoptic meteorology at the University of Uppsala. He remained in this position until he retired in 1961. Tor Bergeron died from pancreatic cancer in Stockholm on June 13, 1977.

LUKE HOWARD AND THE CLASSIFICATION OF CLOUDS

It was not until the 20th century that Tor Bergeron, Walter Findeisen, and Alfred Wegener solved the mystery of how clouds form, but everyone had always known that clouds are important. They bring rain or withhold it, passing tantalizingly over parched fields, and they can bring storms. Aristotle thought the Sun's warmth rearranged water on the Earth's surface, changing it to a substance similar to air, which rises through the air. The Roman naturalist Pliny the Elder (Gaius Plinius Secundus, 23–79 C.E.) believed clouds were made from material from the upper air blended with vapors from the ground that make their way into the atmosphere. A theory that was widely held in the 18th and early 19th centuries supposed that warmth from the Sun shapes water droplets into hollow spheres filled with a highly rarefied form of air, allowing them to rise through the ordinary air to form clouds.

If the way clouds form was a mystery, describing them was no easier. No two clouds are exactly alike, so how should an observer describe the sky to a colleague who had not witnessed it? Theophrastus (371 or 370–288 or 287 B.C.E.; see "Theophrastus and Weather Signs" on pages 6–10) wrote of "clouds like fleeces of wool" and "streaks of cloud." Robert Hooke (1635–1703; see "Robert Hooke and the Wheel Barometer" on pages 52–59) kept meticulous records of the weather and recognized the need for a systematic way to name clouds. In 1667 Thomas Sprat (1635–1713) published the first history of the Royal Society, in which he reproduced Hooke's suggested cloud classification. Hooke's descriptions were still very general:

"Let *Cleer* signifie a very cleer Sky without any Clouds or Exhalations. *Checker'd* a cleer Sky, with many great white round Clouds, such as are very usual in Summer. *Hazy,* a Sky that looks whitish, by reason of the thickness of the higher parts of the Air, by some Exhalation not formed by Clouds. *Thick,* a Sky more whitened by a greater company of Vapours. Let *Hairy* signifie a Sky that hath many small, thin and high Exhalations, which resemble locks of hair, or flakes of Hemp or Flax: whose Varieties may be exprest by *straight or curv'd, &c.*"

The great French naturalist Jean-Baptiste de Lamarck (1744–1829), who devoted great effort to classifying invertebrate animals, also turned his attention to cloud classification. In 1799 Lamarck published the first of a series of annual meteorological digests called *Annuaires Météorologiques,* and, in the third volume, in 1801, he proposed five principal cloud types: hazy clouds (*en forme de voile*); massed clouds (*attroupés*); dappled clouds (*pommelés*); broomlike clouds (*en balayeurs*); and grouped clouds (*groupés*). In 1805 he expanded the list to 12 forms, introducing such terms as flocked clouds (*moutonnées*), torn clouds (*en lambeaux*), banded clouds (*en barres*), and running clouds (*en coureurs*).

It was a noble attempt, but suffered from the same fault as all its predecessors: Lamarck was merely describing the appearance of clouds. Although he intended his scheme to group clouds into distinct categories, there was a serious risk that his list of cloud types would continue growing indefinitely. His scheme suffered a further disadvantage in that he chose rather obscure French names that were not likely to be adopted outside France.

Across the Channel in a basement laboratory in a building in Plough Court, off Lombard Street, London, on an evening in December 1802 (though some historians say it was 1803), a young pharmacist and manufacturing chemist by profession, and meteorologist by vocation, delivered a lecture to the Askesian Society. The Askesian Society, founded in 1796, took its name from the Greek word *askesis* meaning "training," reflecting the intention of its members to learn and improve themselves. They met every fortnight on the premises of a pharmaceutical manufacturing company founded by Joseph Gurney Bevan (1753–1814), but under the control of William Allen (1770–1843) since Bevan's retirement. Under the direction of Allen and his partners the company would eventually become Allen and Hanbury's, which in turn became part of Glaxo Wellcome. William Allen was a close friend of the speaker and a founder member of the Askesian Society.

The title of the lecture was "On the Modifications of Clouds," and the speaker was Luke Howard (1772–1864). Howard recognized that classifying clouds simply by their appearance would inevitably lead to a never-ending list of increasingly obscure names. The task would be impracticable, if not impossible. What he proposed instead was that there are just three very basic types of cloud. All other clouds form from these types, by joining together, breaking apart, or in other ways modifying the basic form. That is what Howard meant by modification, and by the end of his lecture it was clear that he had devised a practical scheme for naming clouds scientifically. Indeed, he had done more, because the names he coined were based on Latin words, not English, which meant that all scientists could adopt them easily, regardless of the language they spoke. The cloud classification scheme meteorologists use today (see sidebar below) has developed from the one described in that lecture, and most of Howard's names have remained unaltered.

Howard isolated three basic types of cloud he called cirrus, stratus, and cumulus. *Cirrus* is Latin for "curl," as in a curl of hair. *Stratus* is the past participle of the Latin verb *sternere*, "to strew," suggesting a layer or stratum. *Cumulus* is the Latin word for "heap," suggesting clouds that form heaps. These three types could then become modified to form cirrocumulus, cirrostratus, cumulostratus (now called stratocumulus), and cumulocirrostratus, or *nimbus*—Latin for cloud, but Howard associated it with precipitation. Howard illustrated his lecture with his own watercolor paintings. In the years that followed Howard continued to revise and improve his classification, and he also devised a set of symbols to represent them, as a form of shorthand for meteorologists recording their observations.

MODERN CLOUD CLASSIFICATION

Clouds are classified in two ways, according to the height of the cloud base and according to the appearance of the clouds (which reflects the manner of their formation). The two methods are complementary, because cloud names are linked to the height of the cloud bases.

The categorization of cloud heights classifies clouds as high, middle, and low. The base heights in each category vary with latitude. Cumulonimbus, which may tower to a very great height, is classed as a low cloud, because of the height of its base.

High cloud. Base height: Polar regions 10,000–26,000 feet (3,000–8,000 m); Temperate regions 16,000–43,000 feet (5,000–13,000 m); Tropics 16,000–59,000 feet (5,000–18,000 m). Cloud types: cirrus, cirrostratus, cirrocumulus

Middle cloud. Base height: Polar regions 6,500–13,000 feet (2,000–4,000 m); Temperate regions 6,500–23,000 feet (2,000–7,000 m); Tropics 6,500–26,000 feet (2,000–8,000 m). Cloud types: altocumulus, altostratus, nimbostratus

Low cloud. Base height: Polar regions 0–6,500 feet (0–2,000 m); Temperate regions 0–6,500 feet (0–2,000 m); Tropics 0–6,500 feet (0–2,000 m). Cloud types: stratus, stratocumulus, cumulus, cumulonimbus

Cloud names divide clouds into 10 basic types (genera). These are (in alphabetical order): altocumulus; altostratus; cirrocumulus; cirrostratus; cirrus; cumulonimbus; cumulus; nimbostratus; stratocumulus; and stratus.

The genera are divided into 14 species: calvus; capillatus; castellanus; congestus; fibratus; floccus; fractus; humilis; lenticularis; mediocris; nebulosus; spissatus; stratiformis; and uncinus.

There are a further nine cloud varieties: duplicatus; intortus; lacunosus; opacus; perlucidus; radiatus; translucidus; undulatus; and vertebratus.

In addition, a smaller accessory cloud may be attached to a main cloud. These include pileus, tuba, and velum. Supplementary features may qualify the description of a main cloud. Such features include arcus, mamma, praecipitatio, and virga.

The modern international classification of clouds is set out, with photographs of all the cloud genera, species, varieties, accessory features, and supplementary features, in the two-volume *International Cloud Atlas,* published by the World Meteorological Organization, 7 bis, avenue de la Paix, CH 1211 Genève, Switzerland (www.wmo.int).

Howard's lecture was serialized in *Philosophical Magazine* in 1803, bringing his ideas to a much wider audience, and his fame spread. It was issued as a pamphlet, and articles on the classification appeared in many other periodicals. It inspired Percy Bysshe Shelley (1792–1822) to write his poem "The Cloud," and Howard corresponded with the German writer Johann Wolfgang von Goethe (1749–1832), who wrote *Howards Ehrengedächtnis* (In Howard's honor), a series of short poems on each of Howard's cloud types, and including

But Howard with his clear mind
The gain of lessons new to all mankind;
That which no hand can reach, no hand can clasp
He first has gained, first held with mental grasp.

Goethe was then a towering literary figure and minister in the government of the grand duke of Saxe-Weimar. He contacted Howard through John Christian Hüttner, an official in the German department of the Foreign Office in London, who wrote to tell Howard of Goethe's interest and of the verses, which by then had been published. Goethe wished to know as much about Howard as Howard was willing to tell him. Luke Howard had become a superstar and celebrities were keen to meet him. On July 22, 1813, he visited Jane Austen (1775–1817) at her home in Hampshire.

Luke Howard was born in London on November 28, 1772, the son of Robert (1738–1812) and Elizabeth (1742–1816) Howard. Robert was a successful businessman and a devout Quaker. Luke attended the Hillside Academy, Thomas Huntley's Quaker school at Burford, near Oxford, from 1780 until 1788. Always interested in natural history, Luke's enthusiasm for meteorology was born in 1783, when two major volcanic eruptions, in Iceland and Japan, blanketed the *upper atmosphere* with dust, producing spectacular dawns and sunsets, but also haze and fog, oppressive heat, and a strange, unpleasant, sulfurous smell. The conditions made many people ill. Then on August 18, Luke saw a meteor cross the sky. Luke spent hours at the window, gazing at the sky. It was then that he began to record his meteorological observations, a habit he maintained for more than 30 years. The habit of gazing at the sky remained with him until he died.

After Luke left school, his father sent him as an apprentice to one of his Quaker friends, Ollive Sims, who was a chemist and druggist in Stockport, Cheshire, in the northwest of England. His apprenticeship completed, in 1794 Luke returned to London. He worked for a firm of wholesale druggists, but after a time he managed to borrow enough money from his father to start a pharmacy business of his own. The work was hard and the hours long, but Luke also attended evening classes in chemistry. Through the people he met there, he was drawn into the community of young, highly intelligent, ambitious Dissenters (Nonconformists) like himself. He became friendly with William Allen and joined the Askesian Society. The illustration shows a portrait of Luke Howard painted in about 1807, when he was 35 years old and already famous.

Luke Howard (1772–1864), the manufacturing chemist, pharmacist, and amateur meteorologist who devised the system of cloud classification that formed the basis of the international scheme used today. The portrait is by John Opie (1761–1807). *(Science Museum/Science & Society Picture Library)*

In 1807 Howard began to compile the "Meteorological Register," which was published regularly over several years in *Athenaeum Magazine.* In 1818–19, he published *The Climate of London,* a two-volume work that he republished as an expanded, three-volume edition in 1833. It was the first book ever written on urban climates and contained the first reference to the *heat island* effect. A heat island is an urban area that is warmer than the surrounding countryside. In 1837 he published a series of lectures he had delivered in 1817, as *Seven Lectures in Meteorology.* This was the first textbook on meteorology. *Barometrographia,* his last book, appeared in 1847. Howard was elected a fellow of the Royal Society in 1821 and joined the British Meteorological Society (now the Royal Meteorological Society) in 1850.

Luke Howard married Mariabella Eliot (1769–1852) on December 7, 1796. The couple had three daughters and three sons. After Mariabella's death Luke went to live with his son Robert in Tottenham, London, where he died on March 21, 1864.

BENJAMIN FRANKLIN AND HIS KITE

There are very few scientific experiments that are so spectacular and so famous that they become the stuff of legend. One of those legendary experiments took place during a storm in June 1752 in a field near Philadelphia, when Benjamin Franklin (1706–90) demonstrated conclusively that lightning is an electrical phenomenon. He used for his experiment a homemade kite and a key.

Franklin was very interested in electricity, a mysterious natural force that many scientists were investigating in the 18th century. The investigators generated electricity by means of machines that used friction to produce a static charge. It is the same principle by which an inflated rubber balloon acquires a static charge if it is stroked several times in the same direction across a woolen pullover, but the machines they used were large and able to produce a powerful charge. The charge quickly dissipates, but in 1745 a German physicist, Ewald Jürgen Georg von Kleist (1700–48), invented a device for storing a static charge, and in 1746 a Dutch scientist, Pieter van Musschenbroek (1692–1761), developed this further. Van Musschenbroek worked at the University of Leiden (then called Leyden) and von Kleist may also have worked there at the same time. The device van Musschenbroek is credited with having invented is known as the Leyden jar. The Leyden jar is a capacitor, which is a device that stores a charge. It consists of two electrical conductors separated by an insulator. The Leyden jar was a glass jar (the insulator) lined on both the inside and outside by a layer of metal foil (the conductors). The jar was sealed by a cork through which there was a rod, and there was a metal chain on the end of the rod inside the jar. The outer layer of foil was connected to the ground. When current from a static charge passed through the rod, it was carried along the chain to the inside conductor, and the jar stored it. A person holding a hand toward the rod would experience an electric shock. Leyden jars were able to store quite large charges, and the shock could be fairly violent. If a metal object were brought close to the rod a spark would flash toward it from the rod, often with an audible crackling sound.

Franklin experimented with Leyden jars, and it occurred to him that the spark and crackling sound were rather like a miniature flash of lightning and crack of thunder. Could it be, then, that during a thunderstorm the sky and the ground became a giant Leyden jar, with the water in the cloud and the ground acting as conductors and the air between them as an insulator? It should be possible to test that idea if he could find a way to hold a metal object close to the equivalent of the rod in a Leyden jar. That is how he came to perform his kite experiment.

Franklin never published an account of this experiment. The story that has become legendary has grown from the one written by Joseph Priestley (1733–1804; see "Karl Scheele, Joseph Priestley, and Dephlogisticated Air" on pages 14–19) in his book *History and Present Status of Electricity,* published in 1767. There were no independent eyewitnesses to the experiment, however, so Priestley must have learned of it directly from Franklin, and Franklin read Priestley's account before it was published. Priestley's version of the story is, therefore, almost certainly true. It would be quite out of character for either man to have fabricated the account. On October 19, 1752, in a letter to his friend, the scientist Peter Collinson (1694–1768), Franklin described how to construct the kite and perform the experiment.

The illustration that follows shows the traditional version of the story. Franklin holds the string of the kite, his young son is beside him, and both of them are standing in the open, beneath a threatening sky. That is not quite what happened. Franklin did not stand beneath a storm cloud flying his kite in the hope of attracting lightning. That would have killed him, as he knew very well.

According to Priestley, Franklin had waited for some time for a spire to be erected on Christ Church in Philadelphia, because he believed the spire would attract lightning, confirming his hypothesis, but then it struck him that a metal rod fixed to a kite would give him access to the electrified region of a cloud. First he made the kite by fixing a large silk handkerchief to a cross, made from two light strips of cedar. He attached a sharp, pointed wire to the top of the upright stick of the cross, extending rather more than 1 foot (30 cm) from the kite. Franklin fitted the kite with a tail and a string made from hempen twine. He attached a loop of silk at the bottom of the string and fastened a metal key to the string.

Franklin found an open field containing a shed, and on a day when a storm threatened he prepared the experiment, assisted by his son William (1731–1813), who was then 21 years old and not a child as the illustration suggests. Between them they raised the kite, then passed the string through the window of the shed. Standing inside the shed, the two men flew the kite from the window toward a promising dark cloud, which passed overhead with no effect. They waited a long time under the lowering sky and were on the point of giving up when Franklin noticed that loose fibers from the string were standing erect and avoiding one another. At once he held his knuckle toward the key

Benjamin Franklin (1706–90) with his son William using a kite and key during a storm in June 1752 to demonstrate that lightning is an electrical phenomenon. *(2003 Getty Images Hulton Archive/Staff)*

and saw a very definite spark. After a short time, when the rain had thoroughly wetted the string, he was able to collect what Priestley called "electric fire" very easily. Franklin also attached the key to the rod on a Leyden jar, collecting charge in that way.

Unknown to Franklin, on May 10, 1752, the French physicist Thomas-François d'Alibard (1703–99) had performed a very similar experiment, using an iron rod 40 feet (12 m) long rather than a kite. D'Alibard also produced sparks from a storm cloud, establishing the electrical nature of the phenomenon.

Franklin's own instructions for the experiment provide more detail. He emphasized the importance of the experimenter being under cover, preferably inside a building with the kite flying from a window or door. This is to ensure that the silk loop remains dry. He held the kite by the silk loop, not the string, and the silk insulated his hand, but only for so long as it remained dry, because water is a good electrical conductor. He wrote that it is essential that the string does not touch the frame of the door or window.

On no account, however, should anyone ever attempt this experiment, because it is extremely dangerous. On August 6, 1753, Georg Wilhelm Richmann (1711–53), a German scientist living in Russia, was killed performing the experiment in St. Petersburg. Other experimenters were also electrocuted while attempting it.

It was this experiment that led Franklin to invent the lightning rod. He tested his design on his own house and later in 1752 lightning rods were fitted to the Academy of Philadelphia (which became the University of Pennsylvania) and Pennsylvania State House (which became Independence Hall).

In Franklin's day it was very difficult for scientists to investigate the interior of clouds. Today, there are aircraft equipped with specialized instruments that fly through storms collecting data. Balloons also take measurements, and meteorological researchers are able to use various types of radar to measure the density of water droplets and the direction and speed of air currents. Although much is still to be learned, their investigations have revealed the way electrical charge separates inside a large cumulonimbus cloud (see sidebar [overleaf]).

During the second half of the 18th century Benjamin Franklin became one of the most famous and admired men in the world. The French economist Anne-Robert-Jacques Turgot (1727–81) wrote of him that: "He snatched the lightning from the skies and the scepter from tyrants." Franklin's enthusiasm for science and his moral views were revealed in a letter he wrote in 1780 to his friend Joseph Priestley: "The rapid progress true science now makes occasions my regretting sometimes that I was born too soon. It is impossible to imagine the height to which may be carried, in a thousand years, the power of man over matter.... O that moral science were in as fair a way of improvement, that men would cease to be wolves to one another, and that human beings would at length learn what they now improperly call humanity!"

Benjamin Franklin was born on January 17, 1706, in Boston, Massachusetts. His father, Josiah (1657–1745), a soap and candlemaker, had emigrated from Banbury, Oxfordshire, in England. Josiah's first wife, Anne Child, died in Boston in 1689, and he married again some months later. His second wife was Abiah Folger, of Nantucket, Massachusetts. Benjamin was Abiah's eighth child and the 15th of Josiah's 17 children. The family could not afford a college education,

CHARGE SEPARATION IN STORM CLOUDS

Lightning is an electrical discharge between regions of opposite charge. The regions of opposite charge may be located within the same cloud, in two separate clouds, in a cloud and the air outside the cloud, or in the cloud and on the ground beneath the cloud. A lightning flash heats the air it contacts by up to 54,000°F (30,000°C) in less than a second. This rapid heating makes the air expand explosively. Thunder is the sound of the explosion.

Electrical charge becomes separated inside a large cumulonimbus storm cloud. Atmospheric scientists are not entirely certain how the charge separates, but they do know that small ice particles and liquid water droplets acquire positive charge, and larger ice particles and droplets acquire negative charge. Once the charge has separated, therefore, it is gravity that carries the particles and droplets with negative charge downward, while air currents carry those with positive charge aloft. The following illustration shows the final arrangement, where positive charge has accumulated near the top of the cloud, negative charge has become concentrated in the lower part of the cloud, and the negative charge has induced a positive charge on the ground surface below.

The doubt centers on how positive and negative charges become separated in the first place. The most likely explanation is that the charges separate when droplets freeze, as a consequence of the freezing process.

A droplet of water contains both positive and negative ions, as hydrogen H^+ and hydroxyl OH^- ($H_2O \rightarrow H^+ + OH^-$). Being lighter, the H^+ ions are more mobile than the OH^- ions. When a droplet freezes, the more mobile H^+ ions migrate toward the cooler region while the OH^- ions remain in the warmer region. A cloud droplet freezes from the outside in, producing

and he spent only one year at a Boston Latin School. He received some private tuition, but was mainly self-taught.

Benjamin left school when he was 10 and, after working for his father for a time, became apprenticed when he was 12 to his brother James, who was a printer. James founded the first independent newspaper in colonial America, the *New England Courant*. Benjamin ran away when he was 17 to start a new life in Philadelphia. He worked as a printer and after a few months sailed for London, returning to Boston in 1726. By 1730 he had his own printing house and published *The Pennsylvania Gazette*, a newspaper that gave him a platform for his ideas.

In the years that followed Franklin became a very successful author, satirist, and political theorist. He was also a statesman and diplomat and, of course, one of the Founding Fathers of the United States. A keen

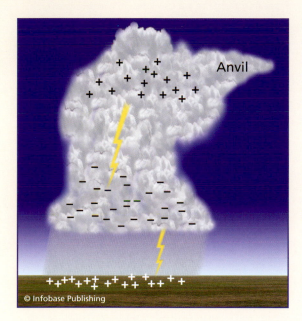

Electric charge has become separated inside a cumulonimbus storm cloud. Positive charge has accumulated near the top of the cloud, and negative charge is concentrated in the lower part of the cloud. The negative charge near the cloud base has induced a positive charge on the ground surface below. At the very top of the cloud, near the tropopause, the wind has swept the ice particles forming the cloud into an anvil shape. The "anvil" is a supplementary cloud feature for which the meteorological name is incus.

a shell of ice enriched in H^+ surrounding a liquid core enriched in OH^-. The core then freezes and expands as it does so. The expansion of the core shatters the outer shell of ice. The tiny splinters of ice from the shell move upward, transported by rising air currents, carrying their positive charge with them. The heavier cores move downward through the cloud, carrying their negative charge with them. The separate positive and negative charges accumulate until a lightning stroke—in fact a giant spark—temporarily neutralizes them.

musician, he played the violin, harp, and guitar, and composed music. As well as the lightning rod, he invented bifocal glasses, the Franklin stove, and an odometer for measuring the distance a coach travels. In 1731 he founded the Library Company of Philadelphia, which was the first public lending library in America. He established the first volunteer firefighting company in America in 1736 and printed currency notes for New Jersey, incorporating his own techniques to prevent counterfeiting. At the end of a very long career, Franklin retired to Philadelphia at the age of 79. He died there on April 17, 1790.

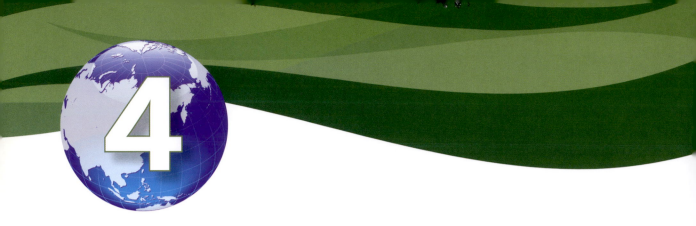

How Gases Behave

The 17th century was a time of vigorous scientific inquiry, and it was also the century during which physical phenomena began to be carefully measured and interpreted mathematically. Galileo (1564–1642; see "Galileo and the Thermometer That Failed" on pages 42–45) emphasized the importance of intellectual rigor, and he exerted a powerful influence on European thought. Measurement, he taught, was the essential precursor to a logical explanation of observations, and logic is highly mathematical. The mathematical approach to science has continued to the present day.

In 1643 Evangelista Torricelli (1608–1647; see "Evangelista Torricelli and the First Barometer" on pages 48–52) showed that air is a material substance that can be weighed and that its weight varies, sometimes in the course of a single day. He also demonstrated the possibility of producing a vacuum. Other scientists explored the reasons for the variability in the weight of air. They found a clear, mathematical relationship between the temperature and volume of air and the pressure it exerts on a vessel containing it. They also discovered the way air pressure—the weight of the atmosphere—changes with increasing altitude.

This chapter describes the work that led to the formulation of the first of the *gas laws*—the formal descriptions of the behavior of gases. The chapter ends with the discovery of *latent heat.* This was a major advance, and one that led to a discovery that was possibly even more important: that heat and temperature are not the same.

ROBERT BOYLE, EDMÉ MARIOTTE, AND THEIR LAW

Robert Boyle (1627–91) was the close friend and patron of Robert Hooke (1635–1703; see "Robert Hooke and the Wheel Barometer" on pages 52–59), and one of the first tasks Boyle set his friend was to make a powerful air pump. In 1657 Boyle read about a pump that the German physicist Otto von Guericke (1602–86) had made, and he asked Hooke to improve on it.

Otto von Guericke had used his pump to perform another of the scientific experiments that have become legendary. The Otto von Guericke Society reproduces the experiment in shows it stages all over the world, von Guericke's original apparatus is held at the Deutsches Museum in Munich, and the experiment has been commemorated on at least two German postage stamps.

Von Guericke made the pump in 1650 and used it for the demonstration that he gave on May 8, 1654, in Regensburg, before an audience comprising members of the government (the Reichstag) and Ferdinand III, the Holy Roman Emperor. In 1656 von Guericke repeated the demonstration in Magdeburg, his hometown where he was the mayor, and that is where his apparatus acquired its name: the Magdeburg hemispheres.

There were two Magdeburg hemispheres, more than 1 foot (30 cm) in diameter. They were made from copper and their rims fitted together snugly, but they were not fixed together. Both hemispheres had a ring to which a chain or rope could be attached, and one hemisphere had a valve that could be sealed to prevent air from entering or leaving. For the experiment, the rims were well coated with grease, and the two hemispheres were joined together tightly. Several people then worked at turning the handle of von Guericke's pump, which removed air from inside the sphere through the valve. When no more air could be extracted the valve was sealed and the pump disconnected. In the 1654 demonstration, von Guericke had 30 horses in two teams each of 15 attached to opposite sides of the sphere. The teams of horses were unable to pull the hemispheres apart, but when the valve was opened, allowing air to enter, the two halves came apart easily. The 1656 demonstration used two teams of eight horses each, and in either 1661 or 1663 von Guericke performed the demonstration again in Berlin for the Elector of Brandenburg, Friedrich Wilhelm, this time with two teams of 12 horses.

Inspired by Torricelli's discovery of air pressure, von Guericke had demonstrated just how powerful that pressure is. As long as the joined hemispheres contained air, the air pressure was equal inside and outside, and the hemispheres came apart easily. When the air was evacuated, the air pressure on the outside of the sphere was not balanced by pressure on the inside, and the hemispheres were held together very tightly. A modern version of this apparatus is still used to demonstrate the force of air pressure. If the hemispheres are 1 foot (30 cm) in diameter and the atmospheric pressure is 14.7 pounds per square inch (1 MPa), the air exerts more than 3 tons (2.7 tonnes) pressure pushing the hemispheres together.

This is the pump Boyle wanted to use for his own experiments, and Hooke made him an improved version. It took time, but the pump was ready by 1659. Boyle did not wait for the pump before commencing his researches, and he published the results of his experiments in 1660, with the title *New Experiments Physico-Mechanical, Touching the Spring of Air and Its Effects.* By the "spring" of air Boyle meant its compressibility. He had made two important discoveries. One was that the weight of a body varies with the density of the air around it, because the air provides buoyancy. His other discovery—not made with the air pump—was that the volume occupied by a gas is inversely proportional to the pressure under which the gas is held. Boyle had found that when he poured mercury into a J-shaped tube that was sealed at the shorter end, the volume of air in the tube decreased, and the volume continued to decrease the more mercury he added. This demonstrated the ease with which air can be compressed—its "spring"—and the relationship can be expressed as: $pV =$ a constant, where p is pressure and V is volume. In English-speaking countries this relationship is known as Boyle's law.

In 1676 the French physicist Edmé Mariotte (ca. 1620–84) published the same law, which he had discovered independently. Mariotte had also observed that a gas expands when it is heated and contracts when its temperature falls. Expansion and contraction obviously alter the volume of the gas and, therefore, $pV =$ a constant is true only if the temperature remains constant. This improved version is known in French-speaking countries as Mariotte's law.

Robert Boyle was the son of Richard Boyle, the earl of Cork. He was born on January 25, 1627, at Lismore Castle, in Ireland. He learned French, Greek, and Latin while still a small boy, and at the age of eight, following his mother's death, he was sent to Eton College, near London. He spent three years at Eton and in 1638, accompanied by a French tutor, he made a tour of Europe, arriving in Italy in 1641. He spent that winter in Florence studying the work of Galileo and returned to England in 1644.

Upon his return Boyle dedicated himself to a life of scientific investigation. Often working with Hooke, he studied combustion, and in 1661 he published *The Sceptical Chymist,* a book describing his chemical experiments, in which he suggested that matter is made up of "corpuscles." These are of varying shapes and sizes, and they join together to form groups, the groups constituting chemical compounds. He was the first scientist to use the word "analysis" to describe the separation of a compound into its constituents. He invented a hydrometer for measuring the density of liquids, and he made the first match, by coating a rough paper with phosphorus and placing a drop of sulfur on the tip of a small stick. The sulfur ignited when the stick was drawn across the phosphorus. He invented a portable camera obscura that could be lengthened or shortened like a telescope to focus an image onto a piece of paper stretched across the back of the box, opposite the lens.

Boyle was drawn into the company of others who shared his interests. These scientists called themselves the "Invisible College" and they used to meet at Gresham College, an institution that still exists. The Lord Mayor of London is its president, and it employs professors to give free public lectures. Some of the members also met in Oxford, and in 1654 Boyle moved to Oxford, which is where he performed his most important work. On August 13, 1662, a charter granted by King Charles II changed the Invisible College into the "Royal Society, for the improvement of natural knowledge by Experiment." The charter named Robert Boyle as a member of the Society's council, and in 1680 he was elected its president, but declined because his conscience forbade him from taking the necessary oath. Robert Boyle was deeply religious and learned Hebrew, Syriac, and Greek in order to be able to read scriptural

Robert Boyle (1627–91), the Irish-English physicist and chemist who discovered the law bearing his name. The portrait, in oil on canvas, was painted in about 1689 by Johann Kerseboom. *(The Granger Collection)*

texts in their original languages. In his will he left money to fund a series of lectures aiming to prove the validity of the Christian religion against the assertions of all other religions.

In 1688, when he was 61, Boyle returned to London and lived with his sister, Lady Ranelagh. The illustration on page 103 shows a portrait painted in about 1689, one year after he had moved to London. He never married. His health had never been robust. He suffered a stroke in 1670, which left him paralyzed, but from which he slowly recovered. So many people visited him that it became burdensome and from 1689 he withdrew increasingly from public life. He died in London on December 30, 1691.

Edmé Mariotte was born about 1620 in Dijon, a town in eastern France where he spent most of his life, becoming prior of the abbey of St. Martin sous Beaune, near the town. He was a scientist as well as a priest, and when the French Academy of Sciences was founded in 1666 he became one of its first members. He wrote papers on many topics, which were published in the *Histoire et Mémoires de l'Académie.* These included *Nouvelles découvertes touchant la vue* (New discoveries concerning vision, 1668), *Traité des couleurs* (Treatise on colors, 1681), and *De la végétation des plantes* (On the growth of plants, 1679 and 1686). His most important work, however, was on fluids. He wrote *Expériences sur la congélation de l'eau* (Findings on the freezing of water, 1682) and *De la nature de l'air* (On the nature of air, 1676 and 1679). It was in *De la nature de l'air* that Mariotte described the relationship between the pressure and volume of air that Robert Boyle had discovered independently, but with the additional recognition that the temperature must remain constant. Edmé Mariotte died in Paris on May 12, 1684.

JACQUES CHARLES AND HIS LAW

In 1699 Guillaume Amontons (1663–1705; see "Guillaume Amontons and the Hygrometer" on pages 68–73) discovered the relationship that came to be known as *Amontons's law.* This states that for any given change in temperature the pressure exerted by a gas always changes by the same amount. It can be expressed as: $P_1T_1 = P_2T_2$, where P_1 and P_2 are the initial and altered pressures, respectively, and T_1 and T_2 are the initial and altered temperatures. A talented inventor, the thermometer he made allowed Amontons to advance beyond the point Mariotte had reached. There was still no standard calibration that would allow him to measure the extent of the change in temperature, however.

It was not until about 1787 that the French physicist and mathematician Jacques Charles (1746–1823) took that next step. Charles was able to measure the amount of change in temperature. Amontons had described the relationship between temperature and pressure; Charles measured the change in volume. He repeated Amontons's work using oxygen, hydrogen, and nitrogen and discovered that for every 1.8°F (1°C) rise in temperature, the volume of the gas increased by 1/273 of the volume it had had at a temperature of 32°F (0°C) and as the temperature fell the volume decreased at the same rate. This is Charles's law, and it can be written as: $V_1/T_1 = V_2/T_2$, where V_1 and T_1 are the initial volume and temperature and V_2 and T_2 are the final volume and temperature, or as $V/T =$ a constant.

Charles had discovered what is now recognized as the second of the gas laws, and one with profound implications. Since with every 1.8°F (1°C) fall in temperature the volume of a gas decreases by 1/273 of its volume at 32°F (0°C), when its temperature falls to -459.4°F (-273°C) its volume must reach zero. This temperature is now called absolute zero. It is the lowest possible temperature, and the third law of thermodynamics states that it can never be reached.

What is known outside France as Charles's law is called Gay-Lussac's law in France. That is because Charles did not publish the results of his experiments, but he did tell Joseph Gay-Lussac (1778–1850) about them. Gay-Lussac repeated the experiments, produced more accurate results, and published them. John Dalton (1766–1844; see "John Dalton and Water Vapor" on pages 26–34) also discovered the relationship, but Gay-Lussac was the first to publish.

Jacques-Alexandre-César Charles was born on November 12, 1746, at Beaugency-sur-Loire in the Loire Valley of France. His education included only basic mathematics and very little science. As a young man he moved to Paris and became a clerk in the Ministry of Finance. It was while he was there that his interest in science was aroused. Benjamin Franklin (1706–90) visited Paris in 1779 as an ambassador for the United States of America, and both he and his scientific work attracted much attention. Charles learned about Franklin's experiments with electricity and began to study experimental physics. By 1781 he was giving public lectures popularizing Franklin's work and demonstrating the experiments with apparatus he made himself. His lectures were popular, and Charles was being noticed.

In June 1783 the Montgolfier brothers made their first experiments with unmanned hot-air balloons at Viadalon-les-Annonay in the south of France (see sidebar "History of Ballooning" on page 139). When news of this reached Paris, Charles began experimenting and found that hydrogen was a better lifting gas than hot air. In August 1783 he and his friends the brothers Anne-Jean (1758–1820) and Marie-Noël (1760–1820) Robert launched the first unmanned flight of a hydrogen balloon, and on December 1, 1783, Charles and Marie-Noël made the first manned flight in a hydrogen balloon. His balloon flights made Charles famous—much more famous than his discovery of the gas law—and the king, Louis XVI, invited him to move his laboratory to the Louvre.

Charles was elected to the French Academy of Sciences in 1785 and later became professor of physics at the Conservatoire des Arts et Métiers. He died in Paris on April 7, 1823.

BLAISE PASCAL AND THE CHANGE OF PRESSURE WITH HEIGHT

Evangelista Torricelli (1608–47) had demonstrated in 1643 that air has weight, and many scientists sought to repeat Torricelli's experiment and improve on his barometer. One of these was Blaise Pascal (1623–62), a brilliant young French physicist, mathematician, inventor, and philosopher. Pascal repeated the experiment, but used red wine mixed with water rather than mercury. Alcohol is less dense than water and much less dense than mercury, so Pascal's barometer tubes had to be 39 feet (12 m) long. They were fastened to the masts of ships. But the low density of the mixture meant that changes in atmospheric pressure produced large movements in the level of liquid in the tube.

Pascal also pondered the implications of Torricelli's discovery. He reasoned that the atmosphere must blanket the Earth like an ocean of air and, if this were so, then the atmosphere must have an upper boundary, equivalent to the surface of the ocean. That being so, with increasing distance from the surface, the amount of overlying air diminishes. Consequently, the weight of the overlying atmosphere must decrease with increasing altitude and so must the atmospheric pressure registered by a barometer. Water cannot be compressed, however, and the ocean surface is clearly defined. Air is easily compressed, so compression by the weight of overlying air must mean its density is at a maximum near the land surface and with increasing altitude it should become increasingly tenuous. Eventually the air must be so thin as to be barely present at all. The atmosphere must have an upper boundary, but not a boundary that is clearly defined.

At the time, Pascal was living in Paris, but in 1646 he returned to his hometown of Clermont (now Clermont-Ferrand), in the Auvergne region of central France, and to a landscape that afforded an opportunity to test his idea. Close to Clermont there are several extinct volcanoes called *puys,* towering to a considerable height and covered with vegetation. If Pascal's ideas were correct, the air pressure at the tops of these conical *puys* should be lower than the pressure at the foot. He set out to conduct a test on the biggest of them, Puy-de-Dôme, 4,806 feet (1,465 m) high.

The sides of Puy-de-Dôme are steep. Today there is a road all the way to the top, but in Pascal's day it meant a hard climb and Pascal was not strong. His health had always been poor; he suffered from headaches, chronic indigestion, and insomnia, and he was partially paralyzed and unable to walk without crutches. It was quite impossible for him to climb Puy-de-Dôme himself, so he enlisted the help of Florin Périer, the husband of his elder sister, Gilberte. It took several months of persuasion, but in the end Florin agreed and on September 19, 1648, accompanied by a small group of friends from Clermont, the two men prepared to take readings of the atmospheric pressure.

The party met at 8 A.M. in a monastery garden that was the lowest point in Clermont and Périer prepared the first of the barometers. These were Torricelli barometers, consisting of glass tubes 4 feet (1.2 m) long and reservoirs of mercury. He measured the pressure in the garden several times and each time the mercury in the tube stood 26 inches (66 cm) above the level of the mercury in the reservoir. He left one of the barometers there, asking one of the brothers to check the barometer from time to time during the day and make a note of any changes in the reading. Périer then set off with another barometer for the top of Puy-de-Dôme, where he checked the pressure in five different places. Each time the mercury reached a height of 23 inches (58 cm). Pascal later repeated the experiment in Paris by carrying a barometer to the top of the bell tower on the church of Saint-Jacques-de-la Boucherie, which was about 165 feet (50 m) high.

Blaise Pascal (1623–62) was the French physicist, mathematician, inventor, and philosopher who discovered that the atmosphere has an upper limit and that atmospheric pressure decreases with increasing altitude. This portrait is from a book published in the 19th century. *(George Bernard/Science Photo Library)*

The experiments established the usefulness of the barometer. They also confirmed Pascal's idea that atmospheric pressure decreases with increasing altitude because there is an upper boundary to the atmosphere and, therefore, the weight of overlying air decreases with increasing altitude. The international unit of pressure, the pascal (Pa), was named in his honor.

Blaise Pascal was born at Clermont on June 19, 1623. His father, Étienne (1588–1651), was a judge and an aristocrat by virtue of his office, and he was interested in mathematics and science. Blaise had two sisters, Gilberte, who was older, and Jacqueline, who was younger. Their mother, Antoinette Bégon, died when Blaise was three. In 1631 Étienne Pascal moved his family to Paris.

Étienne never remarried. He brought up the children by himself and made himself responsible for their education. All three were exceptionally gifted, but Blaise was a mathematical prodigy. At the age of 11 he wrote a short treatise on the mathematics of vibrating bodies and at 12, before he had been taught Euclidean geometry, he worked out a proof that the sum of the angles of a triangle is equal to two right angles. Étienne used to attend meetings of some of the most eminent mathematicians in Europe, which took place in the cell of Père Mersenne (1588–1648), a mathematician, music theorist, theologian, and philosopher who was a mendicant friar belonging to the Order of the Minims. Once Blaise's mathematical talent had become obvious, he was allowed to accompany his father and listen to the discussions. Blaise was especially interested in conic sections, and, in 1640, when he was 16, he wrote *Essai pour les coniques* (Essay on conics) and sent it to Mersenne. It contained what is still known as Pascal's theorem. Mersenne showed it to René Descartes (1596–1650), another member of Mersenne's group, whose supercilious response betrayed his jealousy of the young man.

By that time the family had left Paris. Étienne had disagreed with the fiscal policies of the powerful Cardinal Richelieu (1585–1642) and had been compelled to leave Paris, but in 1639 he was pardoned and made king's commissioner of taxes in the city of Rouen in Normandy. His work was arduous, involving endless calculations of taxes owed and received. The task was complicated by the currency system, in which there were 12 *deniers* in a *sol* and 20 *sols* in a *livre*. (The United Kingdom retained this system until it decimalized its currency in 1971, with the abbreviations £, s, and d for pounds, shillings, and pence.) To ease his father's burden, in 1642 Blaise designed and constructed the "pascaline," which was a mechanical calculator that performed addition and subtraction. By 1652 Pascal had made a total of 50 of these machines, constantly improving them, but they were too expensive for most ordinary people. The pascaline was not the first calculator, because the German astronomer Wilhelm Schickard (1592–1635) had made one in 1624. Nevertheless, the computer programming language Pascal was named in recognition of this achievement. The illustration on page 108 shows Pascal at about this time.

In the 1650s Pascal became increasingly devout, and in 1656 and 1657 he wrote *Pensées* (Thoughts) on human suffering and religious faith. These were published after his death and became his most famous work. They contain "Pascal's wager" in which he sought to establish a rational basis for belief. "If God does not exist, one will lose nothing by believing in him, while if he does exist, one will lose everything by not believing."

Pascal returned to Paris in 1657 and in 1659 he fell seriously ill. He had never been well and for most of his life he was in fairly constant pain. Now bedridden, he continued to work, but in 1662 his condition deteriorated and he died in Paris on August 19, 1662. The nature of his illness remains something of a mystery. His symptoms suggest he may have suffered from stomach cancer, and an autopsy revealed damage to his brain that may explain his frequent headaches. He was buried in Paris, in the cemetery of the church of Saint-Étienne-du-Mont, near the Panthéon.

JOSEPH BLACK, JEAN-ANDRÉ DELUC, AND LATENT HEAT

Throughout the 18th century chemists were steadily accumulating knowledge about the composition of the atmosphere. One of the most prominent of the chemists, and also a physician, was Joseph

Black (1728–99), at the University of Edinburgh. It was Black who, in 1754, first applied the name "fixed air" to the gas that is now called carbon dioxide. The gas had been isolated earlier by Jan Baptista van Helmont (1577–1644; see "Jan Baptista van Helmont and the Discovery of Gases" on pages 10–14), but Black was the first chemist to isolate it in its pure form. Black discovered that certain natural processes release fixed air and, therefore, that it is a normal constituent of air. That means that air is a mixture of gases that can be separated, rather than a single substance.

Black's study of fixed air was the subject of the thesis he submitted in 1754 for his medical degree, and he expanded on it in 1756, in a book with the title *Experiments Upon Magnesia Alba, Quicklime, and Some Other Alcaline Substances.* Black had found that when he heated magnesia alba (magnesium carbonate, $MgCO_3$) it emitted a gas that was different from ordinary air, and he weighed the material to determine exactly how much of the gas was produced. He then repeated the experiment with other "alcaline substances," including calcium carbonate ($CaCO_3$). The same gas was produced and the calcium carbonate was changed to quicklime (calcium oxide CaO). Quicklime, he discovered, will recombine with the gas, to form calcium carbonate. Using modern symbols, the reaction Black studied was:

$$CaCO_3 \leftrightarrow CaO + CO_2$$

These experiments, like all of Joseph Black's work, were conducted meticulously. Black weighed and measured everything with the greatest care, and his experimental results were accurate and very reliable. It was because the gas he had produced could be "fixed" by removing it from the air and absorbing it in a solid substance that he coined the name "fixed air." When he investigated its properties, Black found that a candle flame was extinguished in fixed air and that mice died in it.

By around 1760 Black was growing interested in a quite different problem, arising from an observation that was possible only because of the care with which he made measurements. When he warmed ice he found that it melted slowly, but as it did so its temperature did not change, despite the fact that heat was being applied to it. He concluded that heat and temperature are different.

Heat can have quantity—there can be an amount of heat—and also intensity. Heat is a form of energy, and temperature is one measure of its effect. Thermometers measure the intensity of heat, but not the amount. While the ice is melting it absorbs a quantity of heat that he supposed must mix with the particles of ice, becoming hidden in their substance. He called this latent heat, latent meaning hidden. In 1761 he was able to verify this experimentally and in April 1762 he mentioned it in a lecture to a Glasgow literary society. In 1764 his assistant, William Irvine (1743–97), helped him measure the much larger amount of latent heat that is absorbed when water boils and released when water vapor condenses. In this case their measurements were not very accurate, but they established the principle.

Black did not publish his work on latent heat, and in 1761 the Swiss geologist, physicist, and meteorologist Jean-André Deluc (1727–1817) made the same discovery independently. Deluc also discovered that water is denser at 40°F (4°C) than it is at any higher or lower temperature.

Joseph Black was born in Bordeaux, France, on April 16, 1728. His father, John Black, was Scottish, but born in Belfast, Northern Ireland. John was a wine merchant and his wife, Margaret Gordon, from Aberdeen, Scotland, was also from a family engaged in the wine trade. In 1740 Joseph was sent to school in Belfast, where he learned Latin and Greek, and in 1744, when he was 16, he enrolled at the University of Glasgow to study art, but his father prevailed on him to study a subject that would lead to a profession, and Joseph chose medicine. As a medical student, Joseph had to study chemistry. His professor was William Cullen (1712–90) and, unusually for a student, Joseph performed laboratory experiments, and his relationship to the professor was more like that of an assistant than a student. Black did not graduate from Glasgow, however, because the medical school at the University of Edinburgh was more prestigious and he moved there in 1751 to complete his studies.

Joseph Black (1728–99), the Scottish chemist who discovered latent heat and was the first to distinguish between heat and temperature. The picture is from a 19th-century book. *(George Bernard/Science Photo Library)*

After qualifying as a doctor in 1754, Black began to practice as a physician. In 1756 he returned to Glasgow, to succeed William Cullen as professor of chemistry. He was also appointed professor of anatomy at the University of Edinburgh, but soon changed to professor of medicine. In 1766 Black became professor of chemistry at Edinburgh, once more succeeding Cullen, who had taken up another medical professorship. Typically, Black delivered five lectures each week. The portrait on page 111 shows Black at the height of his career. His position as professor of chemistry was unsalaried, and students paid him to attend his lectures. He combined his research and teaching with his very successful and popular medical practice. Joseph Black never married. He died in Edinburgh on November 10, 1799.

Jean-André Deluc was born in Geneva on February 8, 1727. His father, François Deluc, taught him privately, concentrating on mathematics and science. His education completed, Jean-André entered commerce, where he worked until he was 46. In 1773, his business interests having suffered reverses, he moved to England where he had more freedom to pursue his scientific interests. He had already written on geology and other aspects of natural history and in 1773 he was elected a fellow of the Royal Society and appointed reader to Queen Charlotte (1744–1818). He held this post for 44 years. It brought him an income, but made few demands on his time.

In addition to his other discoveries, Deluc was the first person to calculate the mathematical relationship between air pressure and altitude. He showed that an increase in altitude is proportional to a decrease in the logarithm of air pressure, and inversely proportional to air temperature. Jean-André Deluc died at Windsor, near London, on November 7, 1817.

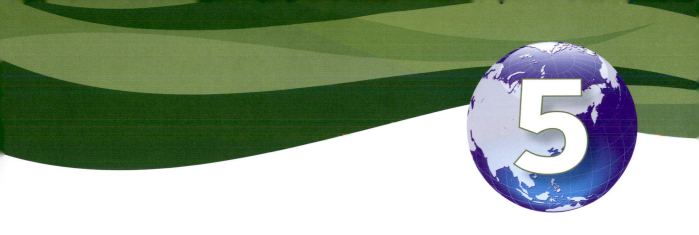

How Air Moves

On October 18, 1938, the S.V. (sailing vessel) *Moshulu* set sail from Belfast, Northern Ireland, bound for Port Lincoln, South Australia, sailing in ballast (without a cargo). She reached Port Lincoln on January 8, 1939, and sailed again, from nearby Port Victoria, on March 11 with a cargo of grain and reached Queenstown, Ireland, on June 10, 1939. It was a round trip of approximately 30,000 miles (48,270 km) that had lasted for one week less than eight months, and with this voyage the *Moshulu* won the last-ever grain race.

The *Moshulu* was a sailing ship—one of the last. Technically, she was a four-masted, square-rigged barque, and she was a giant, built from steel, and 320 feet (98 m) long from stem to stern, and 47 feet (14 m) wide. The tallest of her steel masts was 198 feet (60 m) high, and sailors were required to climb to the top of it with the ship moving beneath them. As the *Moshulu* and the 12 other ships racing against her crossed the ocean they were driven only by the wind. Moving air was their sole source of power, and the *Moshulu* had 31 canvas sails to catch it, the largest of them weighing about 1.7 tons (1.5 tonnes).

The grain races were not sport. In those days Britain imported large amounts of bread wheat from Australia, and the first cargo to unload sold for the best price. Each grain ship carried about 4,500 tons (4,000 tonnes) of grain. It took six weeks to load. As late as the

1930s, sailing ships could still travel out to Australia empty, load, return with a cargo, and show a profit on the voyage. But within a few years steamships had replaced them.

In 1819 the *Savannah* crossed the Atlantic from Savannah, Georgia, to Ireland using steam engines to augment its sails—but made the return voyage under sail alone. More steamships made ocean crossings in the years that followed, but it was not until the middle of the 20th century that big sailing ships ceased to be economically viable. Many of the older ships were sunk during the two world wars, and it was not worthwhile to replace them. Now only a few survive, built more recently and used for training and recreation. Sailing ships had had a very long history. The Egyptians were using sail in about 4000 B.C.E., so for 6,000 years those who wished or needed to cross the sea had no choice but to travel under sail. That means they relied on the wind, and if the wind failed, leaving their ship becalmed, they risked exhausting their supply of drinking water. Mariners staked their lives on their understanding of winds.

Wind has always been economically important. Not only did it power ships, making international trade possible, it also drove vital machinery. Farmers took their grain to a windmill to be ground into flour. Today wind energy is being harnessed anew, this time to generate electricity.

This chapter tells how scientists sought to explain the winds. The first puzzle they addressed centered on the *trade winds* that blow on either side of the equator. These are the most reliable of all wind systems, but why are they so dependable and why do they blow from the northeast in the Northern Hemisphere and from the southeast in the Southern Hemisphere? Winds vary greatly in strength, from the lightest breeze to a hurricane, but if people are to discuss wind strength they need some way to describe it. The first scheme for classifying winds is still used today. This chapter explains how it came into being and how scientists learned that wind direction and strength are related to the distribution of atmospheric pressure and, therefore, how they are linked more generally to weather systems. Finally, the chapter tells the story of the French physicist and highway engineer who discovered why winds and ocean currents flow in circles rather than straight lines.

EDMOND HALLEY, GEORGE HADLEY, AND THE TRADE WINDS

In the language spoken by the Saxons, *trada* meant "footstep" or "track," and by the middle of the 16th century its English meaning had become "path," "way," or "trail." When English sailors first began to cross the equator on the sea route from Europe, around Africa, and to Asia they found that the winds on either side of the equator are very predictable. They almost always blow from the same direction, as though they are following a path—or trade. So they called them the trade winds.

The trade winds blow from the northeast on the northern side of the equator, at an average speed of 11 miles per hour (18 km/hr) and from the southeast on the southern side of the equator at an average speed of 14 miles per hour (22 km/hr). They occur all the way around the world. So far as mariners are concerned, the trades are strongest and most dependable on the eastern side of the Atlantic, Indian, and Pacific Oceans, and their strength and location vary a little with the seasons.

Although their existence had been known for centuries, it was not until 1686 that the English astronomer Edmond (also spelled Edmund) Halley (1656–1742) attempted to explain the reason for them. Nowadays Halley is probably best known for the comet that has borne his name since 1758. The comet had appeared in 1682, and Halley calculated that it was the same comet that had previously been seen in 1531 and 1607. In 1705 he predicted that the comet would return in December 1758. It reappeared in the sky on December 25, 1758, and was given his name. Halley had also been a sailor. He had commanded a warship, the *Paramore Pink*, exploring coasts on either side of the Atlantic, investigating tides, inspecting harbors, and noting variations in compass readings.

Halley's explanation for the trade winds began with the observation that the Earth's surface is heated more strongly at the equator than it is at higher latitudes. Halley suggested that this strong heating causes air to rise by *convection*. Cooler air from higher latitudes on both sides of the equator then flows toward the equator to replace it. His idea was plausible, but incomplete. The mechanism he proposed would produce winds blowing toward the equator from due north and due south. Halley had failed to account for the easterly component of the winds.

That explanation came in 1735, from the English meteorologist George Hadley (1685–1768), in "Concerning the Cause of the General Trade Winds," a paper he presented to the Royal Society. Hadley agreed with Halley that warm air rises by convection over the equator, but he noted that while the cooler air is flowing toward the equator, the Earth itself is also turning on its axis beneath it. Consequently, the flow of air is deflected in a westerly direction, producing winds that blow from the northeasterly and southeasterly directions that are observed.

George Hadley took his explanation further, suggesting a much larger circulation. He proposed that air rising over the equator moves away from the equator at high level and in both directions. When the high-level air reaches the Arctic and Antarctic it subsides and flows back toward the equator at low level. Hadley described a large convection cell in each hemisphere, and he was partly correct. There are atmospheric convection cells similar to those he described, but there are more than two of them. Hadley had overlooked the fact that in middle latitudes the prevailing winds blow from the west and his explanation could not account for the midlatitude westerlies. It was a noble effort, however, and the set of tropical convection cells—several in each hemisphere—are still called Hadley cells. In these, air rises over the equator and moves away from the equator at high level, just as Hadley maintained, but the air descends over the subtropics, not over the poles, and returns to the equator from there, as the trade winds.

At the time, Hadley's paper to the Royal Society aroused little interest. It was not until 1793, long after Hadley had died, that John Dalton (1766–1844; see "John Dalton and Water Vapor" on pages 26–34) recognized its importance. Thanks to Dalton, George Hadley's contribution to meteorology is now acknowledged.

Edmond Halley was born in 1656 at Haggerston, Shoreditch, which was then a village close to London (it is now a district in London's East End). His date of birth was October 29 by the Julian calendar, which was still being used in England. According to the Gregorian calendar, adopted in 1752 and the one now in use, he was born on November 8.

Edmond's father, also called Edmond, was a soap-maker at a time when the use of soap was becoming more widespread, and he became prosperous. He could afford a good education for his son, despite having lost much of his property in the Great Fire of London of 1666. At first he employed a tutor; then he sent Edmond to St. Paul's School, where the boy excelled at Latin, Greek, and mathematics and showed a keen interest in astronomy. He entered Queen's College, University of Oxford, in 1673, at the age of 17, already equipped with a fine set of astronomical instruments provided by his father. While still a student Halley wrote a book on the laws of Johannes Kepler (1571–1630) and published a paper on astronomy in the *Philosophical Transactions of the Royal Society*. John Flamsteed (1646–1719), the Astronomer Royal, recruited Halley as an assistant in 1675, and the following year, with Flamsteed's backing and financial support from the king as well as from his father, Halley sailed to St. Helena, where he spent two years charting the stars of the Southern Hemisphere. On his return to London in 1678, Halley was made a fellow of the Royal Society.

Halley had left university without obtaining a degree, but in 1678 the university awarded him one without requiring him to sit for the examination. In 1679 the Royal Society sent him to Gdansk (then called Danzig) to resolve a dispute between Robert Hooke (1635–1703; see "Robert Hooke and the Wheel Barometer" on pages 52–59) and the German astronomer Johannes Hevelius (1611–87). Hooke had maintained that measurements Hevelius had made without using a telescope could not be accurate. Halley checked them and determined that they were. Halley became friendly with Isaac Newton (1642–1727) and paid for the publication of Newton's greatest work, *Philosophae Naturalis Principia Mathematica* (Mathematical principles of natural philosophy).

Although he was principally an astronomer, in the 17th century astronomers also took part in surveying and mapping the Earth, and Halley studied tides and winds, as well as other meteorological phenomena. In 1704 Halley was appointed Savilian professor of geometry at the University of Oxford and in 1720 he succeeded Flamsteed as Astronomer Royal, a post he held for 21 years. Flamsteed had disliked Newton and when Halley became friendly with Newton, Flamsteed turned against him and their relationship continued to deteriorate over the following years. When Halley became Astronomer Royal, Flamsteed's widow was so angry she had all her husband's instruments removed from the Royal Observatory and sold, so Halley could not use them.

Edmond Halley died at Greenwich, London, on January 14, 1742, by the Julian calendar and January 25 by the Gregorian calendar. The inscription on Halley's tombstone states that he died in 1741. This is correct, because in 17th-century England the year began on March 25, not January 1.

George Hadley was born in London on February 12, 1685. He studied law and became a barrister. Under the English legal system this is an advocate who is permitted to represent clients in the higher courts. He was very interested in atmospheric physics, however, and the Royal Society placed him in charge of the meteorological observations that were collected on its behalf. He held this position for seven years. Hadley died at Flitton, Bedfordshire, on June 28, 1768.

FRANCIS BEAUFORT AND HIS WIND SCALE

One of the tasks of a ship's captain is to maintain a log, or journal, of each voyage. This should record the vessel's location, its progress, every important event that takes place on board, and the weather and sea conditions. The captain must make an entry every day. The log remains with the ship and develops into an account of its history and its crews. The general story is of interest to historians, but the record of weather and sea conditions is of more immediate value. The logs from many ships, written over many years, provide a detailed description of the *climate* in different parts of the world far from land, and until there were orbiting satellites to monitor the oceans constantly, ships' logs were the only source of this information.

To be of real value, however, weather and sea conditions should be described in such a way that accounts written by different captains can be compared easily, which means that every captain must use a similar vocabulary to describe similar conditions. This is not necessarily difficult. "Thick ice on the masts" is fairly self-explanatory, and every seaman recognizes descriptions of the appearance of the sea such as "sea covered with foam," "white foam blown in streaks," and "wave crests breaking into spindrift." In the days before ships carried anemometers to measure wind speed, describing the wind was more difficult, because no matter how hard or gently it blows the wind is always invisible. Its appearance does not change. It is impossible to report what the wind looks like, highly subjective to report what it feels like, and so the only way it can be described is by its visible effects. That is what an anemometer does. It measures the force the wind exerts on a hinged plate or on cups that spin around a vertical axis and translates that force into a speed.

Lacking anemometers, 17th-century captains needed an alternative, and a naval officer, Francis Beaufort (1774–1857), found one that every captain could use without the slightest difficulty. Beaufort instructed captains to report how their sails were deployed. Officers acquired years of sea experience before they were given command of a warship and by then they knew very well how much sail a particular ship should carry under every kind of wind. So all Beaufort had to do was describe winds in terms of the sails a warship should carry, allocate a number to each category of wind, and invite captains simply to record the wind as a number. In 1806 Commander Beaufort as he was then published his *Wind Force Scale and Weather Notation.* He had divided winds into 13 categories according to the force they exerted, from Force 0 (calm) to Force 12 (hurricane). His scale did not mention wind speeds. These would have been impossible to measure accurately and of no practical use at sea (see following sidebar).

The Beaufort wind scale went through several revisions, with Beaufort refining and improving it at every stage. At first it was produced as an aid to captains, which they could use or not as they saw fit. Robert FitzRoy (1805–65; see "Robert FitzRoy and the First Newspaper Weather Forecast" on pages 188–192), the captain of HMS *Beagle,* used it and spoke highly of it. In 1838 the scale was introduced throughout the Royal Navy, and captains were required to use it. In 1874 it was modified once more, this time to include the visible effects of wind on land. With this modification the International Meteorological Committee adopted the scale for use in meteorological telegraphy.

Still the scale made no reference to actual wind speeds. In 1912 the International Commission for Weather Telegraphers began to calculate the wind speeds that would produce the effects described in the scale, but their work was interrupted by the outbreak of World War I. It began again in 1921, when the English meteorologist George Clark Simpson (1878–1965) was asked to undertake the task. His equivalent wind speeds were accepted in 1926. In 1939 the International Meteorological Committee standardized the scale by asserting that the wind speeds are based on values that would be registered by an anemometer set 20 feet (6 m) above the ground. The version of

THE BEAUFORT WIND SCALE

BEAUFORT WIND FORCE SCALE OF 1831

WIND	DESCRIPTION	SAIL
0. Calm		
1. Light air	Or just sufficient to give steerageway.	
2. Light breeze	Or that in which a man-of-war with all sail set, and clean full would go in smooth water from.	
3. Gentle breeze		
4. Moderate breeze		
5. Fresh breeze	Or that to which a well-conditioned man-of-war could just carry in chase, full and by.	Royals, etc. Single-reefed topsails and just carry in chase, full and by, top-gallant sail.
6. Strong breeze		
7. Moderate gale		Double-reefed topsails, jib, etc.
8. Fresh gale		Treble-reefed topsails, etc.
9. Strong gale		Close-reefed topsails and courses.
10. Whole gale	Or that with which she could scarcely bear close-reefed main topsail and reefed foresail.	
11. Storm	Or that which would reduce her to storm staysails.	
12. Hurricane	Or that which no canvas could withstand.	

Note: there are two basic versions of the Beaufort scale. The first is the original one as Beaufort prepared it. The second, set out on page 121, is the one that is used today and that refers to conditions on land. The modern version was subsequently extended to describe hurricanes by the introduction of the Saffir-Simpson hurricane scale, comprising five categories.

the Beaufort scale that resulted has remained in use ever since. It is simple, succinct, and easy to understand.

Francis Beaufort was born on May 7, 1774, in Navan, County Meath, Ireland. The family was descended from Huguenots who had fled from France after the St. Bartholomew's Day Massacre of 1572 and settled in Ireland. Francis's father, Daniel Augustus Beaufort, was a Protestant minister. Daniel was interested in topography—the art of drawing the natural and constructed features of a town or area

BEAUFORT WIND SCALE

FORCE SPEED MPH (km/h)	NAME	DESCRIPTION
0. 0.1 (1.6) or less	Calm	Air feels still. Smoke rises vertically.
1. 1–3 (1.6–4.8)	Light air	Wind vanes and flags do not move, but rising smoke drifts.
2. 4–7 (6.4–11.2)	Light breeze	Drifting smoke indicates the wind direction.
3. 8–12 (12.8–19.3)	Gentle breeze	Leaves rustle, small twigs move, and flags made from lightweight material stir gently.
4. 13–18 (20.9–28.9)	Moderate breeze	Loose leaves and pieces of paper blow about.
5. 19–24 (30.5–38.6)	Fresh breeze	Small trees that are in full leaf sway in the wind.
6. 25–31 (40.2–49.8)	Strong breeze	It becomes difficult to use an open umbrella.
7. 32–38 (51.4–61.1)	Moderate gale	The wind exerts strong pressure on people walking into it.
8. 39–46 (62.7–74)	Fresh gale	Small twigs torn from trees.
9. 47–54 (75.6–86.8)	Strong gale	Chimneys are blown down. Slates and tiles are torn from roofs.
10. 55–63 (88.4–101.3)	Whole gale	Trees are broken or uprooted.
11. 64–75 (102.9–120.6)	Storm	Trees are uprooted and blown some distance. Cars are overturned.
12. more than 75 (120.6)	Hurricane	Devastation is widespread. Buildings are destroyed and many trees are uprooted.

of countryside—and cartography, and in 1792 he produced a map of Ireland. Francis inherited his enthusiasm.

In 1789 Francis left school and went to sea on a merchant ship owned by the British East India Company, but left the company after a year to join the Royal Navy. He was then 16 years old and entered the navy as a cabin boy. Britain was at war with Napoléon, and promotion came quickly. Beaufort was a lieutenant by the age of 22, and, in 1805, aged 31, he was given his first command, of HMS *Woolwich*.

Rear Admiral Sir Francis Beaufort FRS (1774–1857) devised the scale of wind forces that allowed ships' captains to record the winds they experienced in a standard format. This portrait shows Beaufort in about 1850, when he was 76. *(Science Museum/Science and Society Picture Library)*

The *Woolwich* was ordered to survey the sea off the mouth of the Río de la Plata, South America. While he was sailing with the British East India Company, Beaufort had been shipwrecked because of an inaccurate chart, and he understood well the importance of accurate hydrography—surveying of coastlines and the seabed. He surveyed part of the Turkish coast in 1812 on HMS *Frederiksteen,* and, in June 1812, during that expedition, he was seriously wounded while rescuing some of his men who had come under attack from forces commanded by local Turkish rulers. He was taken to a hospital in Portugal, where he made a slow recovery. At the end of the year he was ordered back to Britain. He never went to sea again.

In 1829 Beaufort was appointed head of the hydrographic office of the Admiralty. Under his guidance it became one of the world's finest surveying and charting institutions. He was responsible for the observatories at Greenwich and the Cape of Good Hope, South Africa, which came under Admiralty administration, and he directed several major exploratory expeditions. He was elected a fellow of the Royal Society and became a member of its council, he was a member of the council of the Royal Observatory, and he was one of the founders of the Royal Geographic Society. Beaufort was knighted in 1848, and by the time he retired in 1855—at the age of 81!—he had attained the rank of rear admiral and had served in the Royal Navy for 68 years. The illustration shows him as he appeared in about 1850. After his retirement Sir Francis went to live in Hove, Sussex, where he died on December 17, 1857.

CHRISTOPH BUYS BALLOT AND HIS LAW

In the Northern Hemisphere, if people stand with their backs to the wind there is an area of low atmospheric pressure to their left. This rule-of-thumb makes it possible to form a mental picture of an entire weather system and, from that picture, to make a very approximate weather forecast. Suppose, for instance, that the wind is blowing from the south. People standing with their backs to this wind will have the area of low pressure to their left—in this case to the west. Midlatitude weather systems usually travel from west to east, so the low-pressure center will arrive in the next few hours. The stronger the wind, the lower the central pressure will be, and the lower the central pressure the more likely it is that the weather system will be active, bringing rain or snow depending on the season. It is a simple rule, but a useful one for someone who is out in the open and may need to seek shelter. It does not warn only of approaching bad weather, of course. The rule can also reveal that the low-pressure area has passed and the pressure is rising, promising more settled weather.

Buys Ballot's law states that in the Northern Hemisphere people standing with their backs to the wind have an area of low atmospheric pressure on their left. This is because air flows in a counterclockwise direction around centers of low pressure.

The rule is known as Buys Ballot's law, after the Dutch meteorologist Christoph Buys Ballot (1817–90), who proposed it in 1857, and the illustration shows why it is so. In the Northern Hemisphere air moves in a counterclockwise direction around centers of low pressure and clockwise around centers of high pressure. These directions are reversed in the Southern Hemisphere. Buys Ballot discovered his law by studying many meteorological records going back over many years, searching for patterns that repeat regularly. These revealed that winds always blow in the directions Buys Ballot described around areas of high and low pressure, and not directly toward low-pressure centers and away

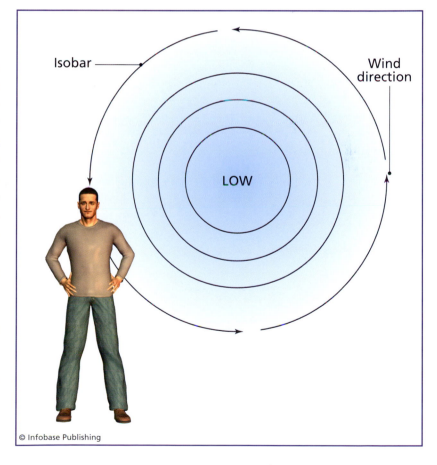

Isobar

Wind direction

LOW

© Infobase Publishing

from high-pressure centers. Some years earlier the French physicist Gaspard de Coriolis (1792–1843; see "Gaspard de Coriolis and Why Air Moves in Circles" on pages 130–135) had shown why this is so, but until then his finding had not been applied to weather systems. In any case, Buys Ballot was aiming to identify patterns, not to explain them.

Buys Ballot published his discovery in *Comptes Rendus,* the journal of the French Academy of Sciences. Unknown to him, a few months earlier the American climatologist William Ferrel (1817–91; see "William Ferrel and Atmospheric Circulation" on pages 125–130) had reached the same conclusion, but by an entirely different route. Ferrel had studied the work of de Coriolis and applied it to pressure systems. It led him to conclude, correctly, that the force acting to move air toward a concentration of low pressure would reach a balance with the Coriolis deflection, and this would result in the air flowing around the low-pressure center, parallel to the *isobars*—lines joining places of equal air pressure. When Buys Ballot learned of this he immediately acknowledged Ferrel's prior claim, but it was too late. His paper had been published, and his name had been attached to the phenomenon. It was called Buys Ballot's law, not Ferrel's law, and the name stuck.

Christoph Buys Ballot also performed one of those experiments that are so dramatic or unusual as to become legendary. In 1845 at Utrecht he demonstrated the Doppler effect. Johann Christian Doppler (1803–53), an Austrian physicist, had discovered that when waves are emitted from a moving source their wavelength is shortened ahead of the source in the direction of travel and lengthened behind the source. This is known as the Doppler effect. It affects light, shortening the wavelength in the direction of the blue part of the spectrum ahead of a moving body and lengthening it toward the red part of the spectrum behind the body. These changes are called a blue shift and red shift, and the pronounced red shift seen in light from distant galaxies proves that the galaxies are receding and, therefore, that the universe is expanding. Meteorologists also measure the Doppler effect using twin radars (Doppler radar) to determine the rate at which the air is rotating inside tornadoes and clouds that are likely to produce tornadoes. Buys Ballot tested the effect, but with sound, not light.

If Doppler was correct, an approaching sound should rise in pitch, because that is what happens when the wavelength of the sound waves is reduced. As the source of the sound passes, the pitch should fall. Today, this is a familiar effect, heard when fast-moving trains, cars, or aircraft pass, but in the 19th century vehicles seldom traveled fast enough for the effect to be clearly discernible. So Buys Ballot recruited a team of professional musicians, whose keen sense of pitch allowed them to recognize the smallest change in a note. He divided them into two groups and stationed one group beside a rail track. The second group, comprising trumpeters with their trumpets, climbed onto an open rail wagon. The trumpeters played a steady note while the train pulled them past the stationary musicians, who noted the rise in pitch as the train approached and fall in pitch after it passed. Buys Ballot had demonstrated the Doppler effect.

Christoph Hendrick Diderik Buys Ballot was born on October 10, 1817, at Kloetinge, Zeeland, in the Netherlands. His father was a clergyman. Christoph was educated at the gymnasium (high school) at Zaltbommel in the south of the country and at the Hogeschool (now university) of Utrecht. He obtained his doctorate in 1844 and was appointed a lecturer in geology and mineralogy at Utrecht, and in 1846 he also began lecturing in chemistry. He became professor of mathematics in 1857, and in 1867 he was appointed professor of physics, the position he held until his retirement. Buys Ballot helped to found the Royal Dutch Meteorological Institute in 1854 and was its first director, holding that position until his death, and in 1873 he became the first chairman of the International Meteorological Committee, the precursor of the World Meteorological Organization. Buys Ballot died at Utrecht on February 3, 1890.

WILLIAM FERREL AND ATMOSPHERIC CIRCULATION

George Hadley had explained the easterly component of the trade winds as the consequence of the Earth's rotation (see "Edmond Halley, George Hadley, and the Trade Winds" on pages 115–118). Everyone found this explanation satisfactory until the middle of the 19th century, when the American climatologist William Ferrel (1817–91) discovered a component in the movement of the air of which Hadley had known nothing. Ferrel found that an additional force was involved. He described this in "An Essay on the Winds and Currents of the Ocean," a paper he published in 1856 in the *Nashville Journal of Medicine and Surgery*. Ferrel wrote:

"In consequence of the atmosphere's revolving on a common axis with that of the earth, each particle is impressed with a centrifugal force, which, being resolved into a vertical and horizontal force, the latter causes it to assume a spheroidal form conforming to the figure of the earth. But, if the rotatory motion of any part of the atmosphere is greater than that of the surface of the earth, or, in other words, if any part of the atmosphere has a relative eastern motion with regard to the earth's surface, this force is increased, and if it has a relative western motion, it is diminished, and this difference gives rise to a disturbing force which prevents the atmosphere being in a state of equilibrium, with a figure conforming to that of the earth's surface, but causes an accumulation of the atmosphere at certain latitudes and a depression at others, and the consequent difference in the pressure of the atmosphere at these latitudes very materially influences its motions."

The force Ferrel described is known today as vorticity, and it arises because a body moving along a curved path conserves its angular momentum, which is the rate at which it turns about an axis, in this case the Earth's axis. Hadley had concentrated on the conservation of linear momentum—the tendency, set down as Isaac Newton's first law of motion, of a moving body to continue moving in the same direction unless acted on by a force.

William Ferrel also constructed a mathematical model of the general circulation of the atmosphere. He published this first in 1856, in the same article in which he corrected Hadley's explanation of the trade winds, and revised it in 1860 and again in 1889. Hadley had proposed a single cell in each hemisphere, but by the 19th century the existence of a polar cell was also recognized. The Hadley and polar cells are both direct cells. That is to say they are driven by convection. In the Hadley cells warm air rises and cool air subsides, forming a convection cell. In the polar cells cold air subsides and, it was assumed, moves beneath warmer air that rises to replace it at high altitude. Ferrel introduced a third set of cells, now known as Ferrel cells, and the general circulation of the atmosphere is now usually described in terms of three cells, known as the three-cell model (see sidebar on page 28). Ferrel deduced their existence from his mathematical studies of the atmosphere. Air from the subsiding side of the Hadley cells moving away from the equator encounters air moving toward the equator from the polar cells. *Convergence* causes the air to rise and divide at high level. Part of the high-level air moves toward the equator until it converges with the subsiding air of the Hadley cells. The subsiding air divides near the surface, part of it flowing away from the equator to complete the Ferrel cells. The Ferrel cells are indirect, in that they are driven by the movement of air in the Hadley and polar cells, and not directly by convection.

William Ferrel was born on January 29, 1817, in Bedford County (now called Fulton County), in southern Pennsylvania. His father, Benjamin, was a farmer and sawmill owner, and William was the eldest of his eight children. In 1829 the family moved to another farm in Berkeley County in what is now West Virginia. William worked on the family farm and for two winters he attended the local one-room school. That was the extent of his formal schooling, but he used to travel to bookstores in Martinsburg and Hagerstown, Maryland, where he purchased science books, which he read avidly. He taught himself so well that he qualified as a schoolteacher. He taught mathematics, and by 1839 he had saved enough money from his salary to enroll at Marshall College (now Franklin and Marshall College) at Mercersburg, Pennsylvania. Ferrel studied there for two years, then, his savings exhausted, returned to teaching until 1842, by which time he had saved enough to complete his studies, this time at Bethany College, in Greater Wheeling, West Virginia, from where he graduated in 1844. After graduating Ferrel returned to teaching, first in Missouri and later in Kentucky. All the time he continued reading books by the world's leading mathematicians and physicists.

Ferrel's interest in the general circulation of the atmosphere was closely linked to his interest in ocean currents and tides, and in 1857 he was invited to join the staff of *The American Ephemeris and Nautical Almanac,* which was published in Cambridge, Massachusetts. Ferrel visited the *Almanac* but returned to his teaching job in Nashville. The following spring, however, Ferrel took up the post and moved to Cambridge. It was while working in Cambridge that Ferrel calculated that winds would move at right angles to the pressure gradient between regions of high and low pressure (see "Christoph Buys Ballot and His Law" on pages 122–125).

On July 1, 1867, William Ferrel was appointed to the United States Coast and Geodetic Survey, where his job was to develop the general theory of tides. At the same time Ferrel widened his research into meteorological phenomena that influence tidal flows. He became increasingly interested in meteorology and wrote on it extensively in the years that followed. His three-volume work *Meteorological Researches* was published between 1877 and 1882, *Popular Essays on the Movements of the Atmosphere* in 1882, *Temperature of the Atmosphere and the Earth's Surface* in 1884, *Recent Advances in Meteorology* in 1886, and *A Popular Treatise on the Winds* in 1889.

THE THREE-CELL MODEL

The movement of air on a global scale transfers heat from the equator to the poles. At the same time, this large-scale movement generates latitudinal belts of prevailing easterly and westerly winds. The movement is known as the general circulation of the atmosphere, and Edmond Halley (1656–1742) and George Hadley (1685–1768) were the first scientists to attempt to describe it as part of their explanations for the trade winds. Nowadays the general circulation of the atmosphere is often depicted as three sets of vertical cells in each hemisphere. This is known as the three-cell model, shown in the illustration (opposite page). It is very approximate, in particular because air moves parallel to the equator as well as at right angles to it; there are several cells in each belt rather than the single cell in each hemisphere shown in the diagram; and over time the cells vary greatly in strength. Nevertheless, the model provides a clear idea of the broad mechanism.

Air, warmed by contact with the land and sea surface, rises over the equator and moves away from the equator close to the *tropopause*. As the air rises its water vapor condenses, producing cloud and the rains of the humid Tropics. The air subsides in the subtropics, at about latitude 30°, dividing as it approaches the surface. The subsiding air has lost most of its moisture, and it heats by compression as it descends, reaching the surface hot and extremely dry and producing the subtropical deserts. Part of the subsiding air returns to the equator at a low level, producing the trade winds. This completes the Hadley cell. The other part of the descending Hadley-cell air moves away from the equator. Very cold, dry air subsides over the North and South Poles and flows away from the poles at low level, producing high-latitude belts of easterly winds. At about latitude 60° N and S the air moving away from the poles encounters air moving toward the poles from the Hadley cells. The converging air rises and divides near the tropopause. Part of the high-level air returns to the poles, where it subsides, completing the polar cells. The remainder of the air moves away from the poles until it meets and joins the subsiding air of the Hadley cells, completing the Ferrel cells.

The trade winds converge at the ITCZ (intertropical convergence zone), which moves with the seasons to the north and south of the equator. The Ferrel and Hadley cells diverge at low level in the horse latitudes, where winds are often light. These earned their name because sailing ships could be becalmed there. Ships often carried cargoes of horses, and if supplies

Ferrel tendered his resignation from Coast Survey on August 9, 1882, in order to accept a position in the Army Signal Service, which at that time ran the weather service. The superintendent of the Coast Survey accepted his resignation, but on condition that he complete the investigations on which he was engaged at the time and he continue to supervise the construction of the machine he had invented to predict tides. His machine worked by a system of levers and pulleys. It

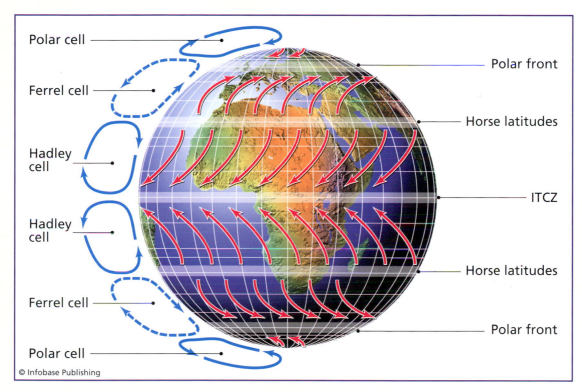

© Infobase Publishing

The three-cell model describes the general circulation of the atmosphere as three sets of vertical cells: the Hadley, Ferrel, and polar cells. The horse latitudes are regions where winds are often calm. The ITCZ is the intertropical convergence zone, where the trade winds converge. The arrows indicate the direction of the prevailing winds.

of drinking water ran low horses sometimes died and their bodies were thrown overboard. The *polar front* marks the boundary between the Ferrel and polar cells. The polar front jet streams blow from west to east in both hemispheres close to the tropopause at the top of the polar fronts. The jet streams produce frontal weather systems that move from west to east.

calculated the combined effects of 19 different factors influencing tides and displayed its results on five dials, showing the times and heights of high and low water at the location for which it had been set. Ferrel had submitted plans of the machine in 1880, construction began in 1881, and the device was completed in 1882. Its first prediction was of the tides for 1885. It remained in use until computers replaced it in 1991.

William Ferrel retired in 1877, when he reached the age of 70. He had never married and went to live with his brother Jacob in Kansas City, but missed his scientist friends and access to good libraries. In 1889 he returned to Martinsburg, where he died on September 18, 1891.

GASPARD DE CORIOLIS AND WHY AIR MOVES IN CIRCLES

Christoph Buys Ballot and William Ferrel, working independently of each other and using entirely different methods, both discovered that air does not flow directly from a region of high pressure to a region of low pressure, but moves around them. Air moves in circles rather than straight lines. Both scientists realized that this was due to the rotation of the Earth, but a French physicist, mathematician, and engineer had already calculated not only how moving air and water are apparently deflected from their straight paths, but also the magnitude of the deflection. Gaspard-Gustave de Coriolis (1792–1843) had described his discovery in "Sur les équations du mouvement relatif des systèmes de corps" (On the equations of relative motion of systems of bodies) in *Journal de l'École Polytechnique* in 1835. The deflection de Coriolis explained was named after him. At first it was regarded as a force and called the Coriolis force, because it appears to push moving air and water to the side. It is not a force, however, but simply an effect, so it is now known as the *Coriolis effect,* but it continues to be abbreviated as *CorF.*

In a sense, the Coriolis effect is an illusion resulting from the fact that people observing it are located on the Earth's surface. The surface is moving, because the Earth is rotating, and, consequently, the observers are moving with it. Air and water are in contact with the surface, but being fluid they are able to move independently of the surface. An observer situated in space who could watch the Earth rotating would see things differently. This observer would see a moving mass of air or water following a straight path, with the Earth moving beneath it. It is only if the path followed by the air or water is drawn on the Earth's surface that it appears to curve. It is apparently deflected to the right in the Northern Hemisphere and to the left in the Southern Hemisphere, as shown in the following illustration.

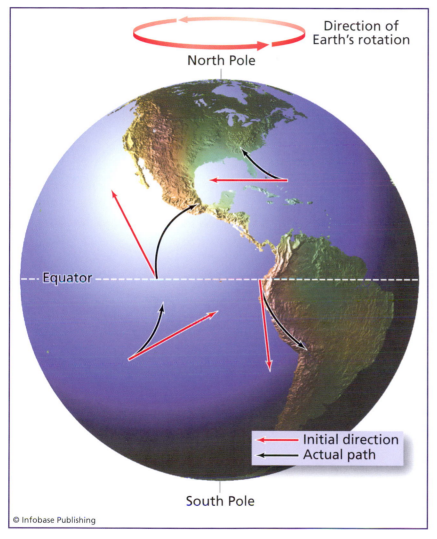

© Infobase Publishing

Moving air masses, winds, and ocean currents that travel long distances across the Earth's surface appear to follow curved paths, deflected to the right in the Northern Hemisphere and to the left in the Southern Hemisphere by an amount that is proportional to the latitude and speed of movement. This deflection was discovered by Gaspard de Coriolis (1792–1843) and is called the Coriolis effect.

The Earth completes one revolution every 24 hours, which means that every point on the surface is moving through 15° every hour in an easterly direction, but because the Earth is approximately spherical, the actual speed varies according to latitude. The circumference of the Earth at the equator is about 24,881 miles (40,034 km), so that is the distance a point on the equator travels in 24 hours. It works out at a speed of about 1,037 miles per hour (1,668 km/hr). A place at 40°N, the approximate latitude of New York City and Madrid, Spain, travels more slowly, because at this latitude the circumference of the Earth is 19,057 miles (30,663 km). New York and Madrid are moving at 794 miles per hour (1,278 km/hr). Suppose an airplane were to take off from a point on the equator due south of New York City and fly due north for the six hours it might take to reach New York. At takeoff the airplane would be traveling eastward (the direction of the Earth's rotation) at 1,037 miles per hour (1,668 km/hr). As it flew north, the airplane would also continue to move eastward at the same speed and after six hours it would be 6,222 miles (10,012 km) to the east of its starting position, as measured on the Earth's surface. During those six hours New York has also moved to the east, but because it travels more slowly it has covered only 4,764 miles (7,665 km). The aircraft will arrive, therefore, not at JFK airport, but 1,458 miles (2,346 km) to the east. That is the Coriolis effect. No force has been exerted, but while the aircraft moved the Earth beneath it also moved, at a different speed.

De Coriolis then calculated the magnitude of this effect. He found that it depends first on the latitude, because of the relationship between the latitude and the speed of motion. This part of the effect is called the Coriolis parameter, and it is equal to $2\Omega \sin \phi$, where Ω is the *angular velocity* (the rate at which a point moves along a curved path, measured in radians per second) and ϕ is the latitude. At the equator the latitude is 0°, and since $\sin 0 = 0$, the Coriolis parameter is zero at the equator. This means there is no Coriolis effect at the equator. The parameter increases with latitude, to the North and South Poles, where the latitude is 90°; $\sin 90 = 1$, so the Coriolis effect reaches a maximum at the poles.

The Coriolis effect also varies according to the velocity (v) of the moving body. This is because the faster a body travels the greater the distance it covers in a given time. When this is taken into account, de Coriolis was able to produce an equation for calculating the magnitude of the effect: $CorF = 2\Omega \sin \phi \times v$.

It was then possible to explain why the wind blows along a circular path around centers of high and low pressure, rather than moving directly from the high-pressure center to the low-pressure center. The difference in pressure between the two centers produces a pressure gradient—like a hillslope, but with high pressure representing high ground and low pressure representing low ground. Isobars—lines joining places of similar air pressure—are equivalent to contours on a topographic map. The pressure gradient exerts a force, called the pressure gradient force (PGF), pushing air from high- to low-pressure centers, like gravity pushing a ball down a hillside. As soon as the air starts to move across the gradient it is subject to the Coriolis effect (CorF), which (in the Northern Hemisphere) deflects the air to the right. A component of the CorF acts in the direction the air is moving. This accelerates the air, but as its velocity increases the CorF increases in proportion, increasing the amount of deflection. When the PGF and CorF reach a balance, the PGF and CorF are acting in opposite directions, with the result that the air moves at right angles to both, and the wind blows parallel to the isobars. The following diagram shows how this balance is reached. If the pressure gradient should become steeper, drawing the wind across the isobars, the CorF would also increase, restoring the balance. A steeper pressure gradient does not make air flow along it, therefore, but accelerates winds that continue to blow parallel to the isobars. The closer together the isobars appear on a weather map, the steeper the pressure gradient is and the stronger are the winds. Although de Coriolis was correct, the situation he described occurs only in air that is well clear of the Earth's surface. That is because the uneven surface causes friction, which slows the wind and reduces the CorF correspondingly, allowing the wind to blow across the isobars at an angle of between 10° and 30°, depending on the wind speed and local topography.

There is also a popular misconception that it is due to the Coriolis effect that water describes a spiral path as it flows out of a bathtub and that it follows a counterclockwise spiral in the Northern Hemisphere and a clockwise spiral in the Southern Hemisphere and that the spiral changes direction if the bath is carried across the equator. The Coriolis effect is much weaker than the other factors affecting the way water flows out of a bathtub, such as the shape of the tub, the

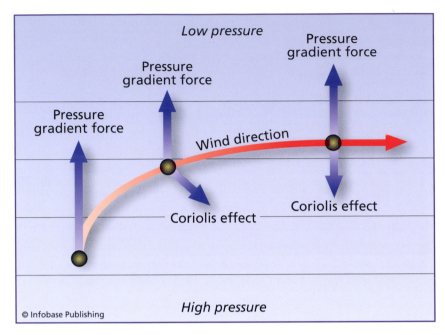

The difference in pressure between the regions of high and low pressure produces a pressure gradient that exerts a force (the pressure gradient force, PGF). As air moves in response to the PGF, it experiences the Coriolis effect (CorF). When the PGF and CorF are in balance, the air flows parallel to the isobars.

shape of the outflow pipe, and the way the water is moving in the tub away from the outflow. In fact, water sometimes flows clockwise and sometimes counterclockwise. In addition, the Coriolis effect does not exist at the equator.

Gaspard-Gustave de Coriolis was born in Paris on May 21, 1792. His family had been lawyers in Provence in southern France and had been made aristocrats in the 16th century, hence the "de" in their name. Gaspard's father, Jean-Baptiste-Elzéar de Coriolis, was an army officer, however, who fought in the American campaign in 1780. The family were stripped of their privileges and wealth following the French Revolution, and in September 1792 they moved to the town of Nancy in eastern France, where Jean-Baptiste-Elzéar became an industrialist. Gaspard spent his childhood in Nancy and that is where he went to school.

In 1808 Gaspard passed the entrance examination and enrolled at the École Polytechnique in Paris, the prestigious school that trained students destined to become government officials. After graduating he entered the École des Ponts et Chaussées (School of Bridges and Highways). While a student there Gaspard spent several years on active service with the corps of engineers, working in the Vosges Mountains and around Meurthe-et-Moselle. Gaspard graduated in highway engineering and was determined to become an engineer, but his health was poor and when his father died Gaspard was left to take care of the family. In 1816 he obtained a post as a tutor in analysis at the École Polytechnique, later becoming assistant professor of analysis and mechanics. In 1829 he was appointed professor of mechanics at the École Centrale des Artes et Manufactures. He held this position until 1836, when he became professor of mechanics at the École des Ponts et Chaussées, and in the same year he was elected a member of the French Academy of Sciences. In 1838 he was made director of studies at the École Polytechnique.

Although de Coriolis is remembered principally for his explanation of the Coriolis effect, he made other important contributions to science. He defined "work" as the displacement of a force through a distance, and he introduced what he called *force vive,* which is now known as kinetic energy, defining it as $\frac{1}{2} mv^2$, where m is mass and v is velocity. In his *Théorie mathématique des effets du jeu de billiard,* published in 1835, he explained the mathematical theory behind the game of billiards—a game similar to pool but played with only three balls. Gaspard-Gustave de Coriolis died in Paris on September 19, 1843.

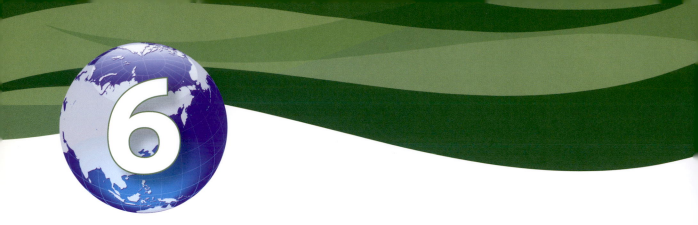

6

Reaching the Sky

Studying the atmosphere is difficult, because most of it is inaccessible. Scientists can collect samples and make measurements at ground level, and they can climb mountains to examine the air at greater elevations, but the air extending for miles above their heads is beyond their reach. That reach was dramatically extended in 1783, when the first hot-air balloon rose into the sky above Annonay, in France (see sidebar "History of Ballooning" on page 139), promising journeys into the sky.

Atmospheric scientists were among the first people to take to the skies in balloons, but they soon realized that crashing was only one of the hazards they faced. The density of the air decreases very rapidly with increasing altitude, and long before a large balloon reaches the maximum altitude of which it is capable, the air is much too thin for humans to breathe. The experience is not unpleasant. The victim often becomes euphoric, but euphoria is quickly followed by confusion and then the loss of consciousness. The experience of early balloonists who climbed too high and felt dizzy gave rise to the expression "the dizzy heights."

The danger of losing control during an ascent was very real, and in 1862 it carried the first balloonists unwillingly into the lower stratosphere. Henry Tracey Coxwell (1819–1900) was a professional aeronaut and very experienced, having made more than 400 flights. He gave many demonstrations. Dr. James Glaisher, FRS (1809–1903), was a meteorologist. For 34 years Glaisher was superintendent of the Department of Meteorology and Magnetism at the Royal Greenwich Observatory. He was also a member of a committee of the British Association for the Advancement of Science and for several years had been urging the Association to sponsor a series of high-altitude balloon ascents in order to study the upper atmosphere. The problem was that such ascents called for a much larger balloon than any balloonist owned. Coxwell agreed to build a suitably large balloon provided the Association would guarantee to pay £50 (about £3,500 in today's money, worth $7,000 at the 2008 conversion rate) for each flight. The Association agreed, Coxwell built the balloon, which he called the *Mammoth,* and the search began for a scientist to volunteer to fly as observer. When no one suitable applied, Glaisher agreed to fly.

The hydrogen-filled *Mammoth* took off on July 17, 1862, from Wolverhampton in the English Midlands and rose rapidly through cloud into a clear sky, with Glaisher checking his instruments and making notes of the temperature, pressure, and humidity, checking the altitude with their aneroid barometer, and recording observations of the cloud formations and wind. On that flight they reached an altitude of about 30,000 feet (9,000 m). Their lips had turned blue and they had difficulty breathing, so Coxwell brought the balloon down. Glaisher found that, rather than decreasing steadily, as they climbed the air temperature changed erratically, sometimes falling rapidly and at one point rising.

They made several more flights and then, on September 5, Coxwell lost control. They took off from Wolverhampton, rose through cloud, and emerged climbing very fast and spinning. In a little more than three-quarters of an hour they were at 26,250 feet (8,000 m) and were still ascending. As they continued rising Glaisher began to have difficulty reading his thermometer, but when he turned to ask Coxwell for assistance he found the aeronaut was out of the basket, on the ring above it, trying to free the rope to the valve that released gas and that had become entangled. At 36,000 feet (11,000 m) Glaisher lost consciousness, and Coxwell, whose hands had frozen to the ring, managed to seize the valve line in his teeth and pull it to release gas, and at last the balloon began to descend. They had reached 37,000 feet (11,278 m). Both men recovered. Glaisher made many more balloon ascents, and in 1866 he became one of the founding members of the Aeronautical Society of Great Britain.

This chapter tells the story of the exploration of the sky. It contains a sidebar with an outline of the early history of ballooning, and an account of the earlier altitude record set by the French chemist and physicist Joseph-Louis Gay-Lussac (1778–1850) in 1804. The chapter ends with an account of the discovery of the stratosphere, and the way it acquired its name.

JOSEPH-LOUIS GAY-LUSSAC AND HIS BALLOON FLIGHT

On December 1, 1783, Jacques Charles (1746–1823) became a popular hero when he and Marie-Noël Robert (1760–1820) made the first ascent in a hydrogen-filled balloon (see sidebar [opposite]). Hydrogen balloons were becoming popular, but almost 20 years passed before they began to be used to take scientific measurements—and to establish the first altitude records. On August 24, 1804, the French chemist Joseph-Louis Gay-Lussac (1778–1850) and the physicist Jean-Baptiste Biot (1774–1862) took to the air from the garden of the Conservatoire des Arts, in Paris, and climbed to 13,120 feet (4,000 m). On September 16 of the same year Gay-Lussac made a solo ascent, this time reaching an altitude of 23,012 feet (7,019 m) above sea level. Balloonists at that time used instruments that tended to overestimate their altitude, so it is possible that Gay-Lussac did not attain quite this height. Nevertheless, he certainly flew higher than the tallest peak in the Alps—Mont Blanc, at 15,775 feet (4,808 m)—and he set an altitude record that remained unbroken for almost 50 years.

Gay-Lussac and Biot had been commissioned by the French Academy of Sciences to investigate the Earth's magnetic field, in particular to discover whether it changed with increasing altitude. Their investigation was possible because the Academy had obtained the use of a balloon. Gay-Lussac decided that the first ascent did not reach a sufficient height and that is why he flew again a few weeks later. On his second ascent he found that the Earth's magnetic field remains constant with increasing height, at least up to the altitude he reached. He also made measurements of the atmosphere. He measured temperature and humidity, and at the top of his climb he recorded an air temperature of 14.9°F (-9.5°C). He also collected air samples, which

HISTORY OF BALLOONING

At a temperature of 32°F (0°C) and pressure of 29.92 inches of mercury (101.32 kPa), 1 cubic foot of air weighs 0.0807 pounds (1 m^3 weighs 1.3 kg). Any gas that is lighter than this can be used to inflate a balloon that will rise through the atmosphere. Hydrogen weighs 0.0055 pounds per cubic foot (0.09 kg per m^3) and helium weighs 0.011 pounds per cubic foot (0.18 kg per m^3). Either of these gases is suitable. Heating a gas causes it to expand, and expansion makes the gas less dense. Consequently, the weight of a given volume of gas decreases as its temperature increases. That is the principle that carried the first balloon aloft, and today the great majority of passenger-carrying balloons use hot air as a lifting gas. That is not because hot-air balloons are inherently superior, however, but because they are more economical and convenient for recreational and sport balloonists.

The first balloon rose on June 4, 1783, reaching a height of 6,560 feet (2,000 m) above the town of Annonay, near Lyons, France. The flight lasted about 10 minutes and the balloon traveled 1.2 miles (2 km). The approximately spherical balloon was made from three layers of paper covered with linen sackcloth and constructed in four sections held together by buttons. A reinforcing net made from cord covered the entire balloon to give added strength. It was 33 feet (10 m) in diameter. When its makers lit a fire of straw and wool beneath it, the bag filled and the balloon took off. The two brothers who built it, Joseph Michel (1740–1810) and Jacques–Étienne (1745–99) Montgolfier, owned a paper mill and knew a great deal about the properties of paper. They knew less about gases, however, believing they had discovered a new, light gas.

News of their success quickly reached Paris, and on September 19 the brothers staged a second demonstration at the royal palace at Versailles, before the king and queen, Louis XVI and Marie-Antoinette. This time Étienne collaborated with Jean-Baptiste Réveillon (1725–1811), a wallpaper manufacturer, and the balloon was made from taffeta coated with an alum varnish. (Alum is hydrated potassium aluminum sulfate ($KAI(SO_4)_2.12H_2O$).) It was painted sky-blue, with golden decorations, and the signs of the zodiac. For this flight the balloon carried a basket with three passengers: a duck, a rooster, and a sheep called Montauciel (Climb-to-the-sky). They were carried to a height of about 1,500 feet (458 m) and landed safely.

The third Montgolfier balloon was much larger and designed to carry human passengers. This balloon was 75 feet (23 m) tall, 46 feet (14 m) across, and held 60,000 cubic feet (1,700 m^3) of air. It was tested in tethered ascents on October 17 and 19, 1783, carrying a young physician, Jean-François Pilâtre de Rozier (1754–85), and on November 21 Pilâtre de Rozier and the marquis d'Arlandes (1742–1809) made the first free, manned flight. The following illustration shows their takeoff in front of a vast crowd from the grounds of the Château de la Muette to the west of Paris. They flew across the city for about 5.6 miles (9 km) and landed 25 minutes later on the Butte-aux-Cailles, just outside the city limits. The fire was carried below the balloon and the two men carried enough fuel for

(continues)

(continued)

a much longer flight. They could not continue, however, because the balloon fabric was catching fire. Sadly, Pilâtre de Rozier and a companion, Pierre Romain, were killed on June 15, 1785, when their hydrogen balloon exploded and crashed in the Pas-de-Calais during an attempt to cross the English Channel. They became the first people to die in an air crash.

On December 1, 1783, Jacques Charles (1746–1823) and Marie-Noël Robert (1760–1820) made the first ascent in a balloon filled with hydrogen, reaching about 2,000 feet (610 m). Charles then made a second ascent alone and reached 8,900 feet (2,700 m). That balloon was made from rubberized silk. Hydrogen balloons then became more popular than hot-air balloons. On January 7, 1785, Jean-Pierre Blanchard (1753–1809) and John Jeffries (1745–1819), an American physician from Boston, made the first crossing of the English Channel in a hydrogen balloon.

On November 26, 2005, Vijaypat Singhania of Mumbai reached 69,800 feet (21,290 m) in a hot-air balloon, setting a world record for hot-air balloons. He took off from Mumbai and landed 150 miles (240 km) away in Panchale. The record

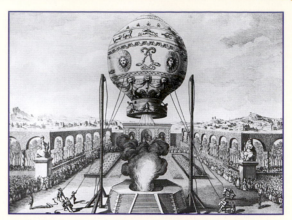

The first-ever manned flight. The ascent of the Montgolfier hot-air balloon on November 21, 1783, carrying Jean-François Pilâtre de Rozier (1754–85) and the marquis d'Arlandes (1742–1809), from the gardens of the Château de la Muette, near Paris. They traveled about 5.6 miles (9 km) before landing 25 minutes later. *(Hulton Archive/Stringer)*

height for a manned hydrogen balloon was set on May 4, 1961, when Commander Malcolm D. Ross (1919–85) and Lt. Cmdr. Victor A. Prather Jr. (1923–61) of the U.S. Navy rose to 113,740 feet (34,668 m). They landed in the Gulf of Mexico and Prather drowned when he fell while climbing the ladder into the rescue helicopter.

he studied on the ground. These revealed that the chemical composition of the atmosphere also remains constant with height up to the altitude he reached. This was the first time a scientist had taken atmospheric samples and measurements at an altitude higher than the tallest European mountain.

Joseph-Louis Gay-Lussac was born on December 6, 1778, at St.-Léonard-de-Noblat, in the district of Haute Vienne, in central France. His father, Antoine Gay, was a judge who added the name Lussac, taken from a small property he owned, to avoid confusion with other families called Gay. Joseph began his education at home with private tutors, but in 1793 his father was arrested on suspicion of show-ing sympathy to the aristocrats and it became difficult to continue teaching the boy at home. The following year Joseph-Louis was sent to Paris to study for the entrance examination to the École Polytech-nique. In 1797 he passed the examination and enrolled at the college. In 1799 he transferred to the École des Ponts et Chaussées (School of Bridges and Highways), where he was assigned to assist the eminent chemist Claude-Louis Berthollet (1748–1822), working at Berthollet's home at Arcueil, near Paris. For a time Gay-Lussac worked alongside Berthollet's son in a linen-bleaching factory. In 1802 Gay-Lussac was appointed a demonstrator to the chemist Antoine-François Fourcroy (1755–1809) at the École Polytechnique.

Gay-Lussac spent 1805 and 1806 on an expedition to measure terrestrial magnetism led by Alexander von Humboldt (1769–1859). From 1808 until 1832 Gay-Lussac was professor of physics at the Sor-bonne University in Paris, and on January 1, 1810 he became profes-sor of chemistry at the École Polytechnique. He resigned his position at the Sorbonne in order to take up the post of professor of chemis-try at the Jardin des Plantes, which was part of the Musée National d'Histoire Naturelle. He was elected to the Academy of Science in 1806. He also held a number of official positions.

Gay-Lussac made his most famous discovery in 1809. Now known as Gay-Lussac's law of combining volumes, this was that when gases react chemically they do so in small, whole-number ratios of their volumes. Working in collaboration with Humboldt, for example, Gay-Lussac found that two parts of hydrogen combine with one part of oxygen to form one part of water. He also discovered that when two parts of carbon monoxide (CO) combine with one part of oxygen they form two parts of carbon dioxide: $2CO + O_2 \rightarrow 2CO_2$.

As well as being a distinguished scientist, Joseph-Louis Gay-Lus-sac was also a politician. He was elected to the Chamber of Depu-ties in 1831 to represent Haute Vienne and in 1839 he entered the Chamber of Peers. He married Geneviève Rojet in 1808. They had five children. Gay-Lussac died in Paris on May 9, 1850.

LÉON TEISSERENC DE BORT AND THE STRATOSPHERE

Although meteorologists were attracted to ballooning in the late 18th century and for most of the 19th century, preferring to carry their instruments aloft and make their own measurements, it was not strictly necessary for them to visit the upper atmosphere in person. There was a simpler, cheaper, and safer way to study the upper air—by instruments carried beneath unmanned balloons made from paper or rubber. The difference was similar to the difference in the early 21st century between manned spaceflights to the Moon and Mars and unmanned missions using robots. The robots are cheaper, because they can tolerate a wider temperature range and do not require food, drink, or air to breathe, but manned missions are more glamorous and more flexible, because human astronauts can seize opportunities and interpret what they see.

Today, balloons are used routinely to monitor conditions in the upper atmosphere. About 500 upper air stations throughout the world release a weather balloon at noon and midnight universal time (the time at Greenwich, England, on the 0° meridian). A weather balloon is about 5 feet (1.5 m) in diameter when fully inflated and is partly filled with helium or hydrogen, which expands as the balloon ascends. Once released, the balloon rises at about 16 feet per second (5 m/s) to an altitude of 66,000–98,000 feet (20–30 km), where it bursts and its package of instruments parachutes to the ground. About one-quarter of the instrument packages are recovered and used again. The instrument package hangs beneath the balloon on a cable 98 feet (30 m) long and contains an aneroid barometer, devices to convert the barometer reading to digital form, a radio transmitter, and batteries for power. Above the main package there is a device that measures humidity and above that a ring supporting the wire of an electrical-resistance thermometer. These instruments also feed data to the radio transmitter. A package of instruments linked to a radio that transmits data to a receiving station on the ground is known as a *radiosonde.*

Kites were used for a time prior to the introduction of weather balloons. In 1749, Alexander Wilson (1714–86) and his student Thomas Meville at the University of Glasgow attached thermometers to a line linking six paper kites that they flew to different heights. Each thermometer was attached to the kite string by a fuse. When the fuse burned through the thermometer fell, paper brushes acting like a parachute. In 1752 Benjamin Franklin (1706–90; see "Benjamin Franklin and His Kite" on pages 94–99) flew a kite to demonstrate that lightning is an electrical phenomenon. In 1893 the American meteorologist Charles F. Marvin (1858–1943), professor of meteorology at the United States Weather Service, and Lawrence Hargrave (1850–1915), an Australian engineer and astronomer, invented a box kite that carried instruments set inside an aluminum cylinder where they recorded data by moving pens on a rotating drum. The instruments recorded temperature, air pressure, humidity, and wind speed. The Marvin-Hargrave kites were first deployed in 1898, and eventually there were 17 stations using them in different parts of the United States. Kites were expensive, however, and because they had to remain tethered the upper atmosphere was beyond their reach.

Balloons came into widespread use in the 1920s, but they were first used in France in the late 19th century, around the same time that Marvin was experimenting with kites in the United States. The meteorologist who pioneered the use of balloons was Léon-Philippe Teisserenc de Bort (1855–1913), who used them to discover that the atmosphere consists of a series of concentric shells (see sidebar on page 144). Teisserenc de Bort (this was his surname) sent instruments aloft with balloons and published his findings in *Comptes Rendus de l'Académie des Sciences,* the journal of the French Academy of Sciences. His experiments confirmed that temperature usually decreases steadily with increasing height, but only to an altitude of about 7 miles (11 km). Above that height he found that the temperature remained constant up to the highest levels his balloons were able to reach. Suspicious of this, he repeated the measurement, eventually sending balloons aloft no fewer than 236 times and always with the same result. At last, in 1902, he published his discovery and its implications.

Teisserenc de Bort reported that the atmosphere was divided into two layers and in the years that followed he developed the idea. Within the lower layer, air is constantly moving, vertically as well as horizontally, and temperature decreases with height. He used the Greek word *tropos,* which means "turning," to describe it. He called it the troposphere. The upper layer was different. There the temperature remained constant, and there was no evidence of vertical motion. That being so, he suggested that the atmospheric gases separate in this layer according to their molecular weights, with the heaviest at the bottom and the lighter gases above. In this region of

STRUCTURE OF THE ATMOSPHERE

The atmosphere envelops the Earth, but its characteristics change with distance from the surface, so its structure is that of a number of concentric spheres. Each of these spheres is known as an atmospheric shell or layer. About half of the mass of the atmosphere is contained in the region below about 3.5 miles (5.5 km) altitude, and 90 percent of the atmosphere lies between the surface and a height of 10 miles (16 km). Above 10 miles (16 km), the atmosphere extends to at least 350 miles (550 km) above the surface, but this part of the atmosphere contains only 10 percent of the total mass. At the top of the atmosphere the air density is about one-trillionth (million-millionth) of the sea-level density.

The lowest atmospheric layer is the *troposphere,* within which air temperature decreases with increasing altitude. This is the layer that contains almost all of the atmosphere's water vapor, and it is the layer in which weather phenomena occur. The tropopause, which is the upper boundary of the troposphere, is a region where temperature ceases to decrease with increasing height. The height of the tropopause averages 10 miles (16 km) over the equator, 7 miles (11 km) in middle latitudes, and 5 miles (8 km) over the poles. The average air temperature at the tropopause is -85°F (-65°C) over the equator, -67°F (-55°C) over middle latitudes, and -22°F (-30°C) over the poles.

All of the atmosphere lying above the tropopause is known as the upper atmosphere. The *stratosphere* extends from the tropopause to the *stratopause,* in summer at a height of about 34 miles (55 km) over the equator and poles and 31 miles (50 km) in middle latitudes. In winter it is at about 30 miles (50 km) over the equator and 37 miles (60 km) over the poles. Temperature remains constant with increasing height in the lower stratosphere. Above about 12 miles (20 km), temperature increases with height, more

the atmosphere the air would form layers, or strata, of different gases, so he called the layer the stratosphere. He was mistaken, and the two atmospheric layers he had identified have the same chemical composition and in both of them the air is thoroughly mixed, but the names he coined have stuck. The two lower layers of the atmosphere are still called the troposphere and stratosphere, separated by a boundary called the tropopause.

The discovery of the layered structure of the atmosphere had widespread implications, because if the atmosphere was layered, was it possible that the oceans and even the solid Earth were also constructed in layers? Between 1902 and 1904 the Swedish oceanographer and physicist Vagn Walfrid Ekman (1874–1954) discovered that the waters of the ocean also form layers. A few years later, in 1909, the Croatian seismologist Andrija Mohorovičić (1857–1936) found rapidly above about 20 miles (32 km), sometimes reaching 32°F (0°C) at the stratopause, where the average atmospheric pressure is 0.69 pounds per square inch (100 Pa). The rise in temperature is due to the absorption of ultraviolet radiation by oxygen (O_2) and ozone (O_3).

The stratopause, a region within which temperature remains constant with increasing height, marks the boundary between the stratosphere and the *mesosphere*. Temperature decreases rapidly with height throughout the mesosphere, to an average -22°F (-30°C) in summer and -130°F (-90°C) in winter at the *mesopause*, which is the upper boundary of the mesosphere at about 50 miles (80 km), above which lies the *thermosphere*. Temperature remains constant with height in the lower thermosphere but increases rapidly above about 55 miles (88 km), reaching up to 1,830°F (1,000°C) at the *thermopause*, 310–620 miles (500–1,000 km) above the surface, the height varying with the amount of ultraviolet radiation that

oxygen atoms absorb. Satellites moving through the thermosphere are not heated by friction with the air, but they are heated by absorbing solar radiation. In the region above the thermopause, known as the *exosphere*, air atoms and molecules are so widely scattered that collisions between them are rare events. The following illustration shows the structure of the atmosphere.

The chemical composition of the atmosphere remains constant throughout the troposphere, stratosphere, and mesosphere. This region of the atmosphere is known as the *homosphere*, above which lies the *heterosphere* in which the atmospheric gases begin to separate. In the lower thermosphere the air consists mainly of molecular nitrogen (N_2), molecular oxygen (O_2), and atomic oxygen (O). Above about 125 miles (200 km), O is more common than N_2 or atomic nitrogen (N). In the upper thermosphere the predominant gas is atomic helium (He), and above that atomic hydrogen (H).

that the Earth's crust and mantle are also distinct layers, separated by what is now known as the Mohorovicic discontinuity, or Moho.

Léon-Philippe Teisserenc de Bort was born in Paris on November 5, 1855. His father had been an engineer and at the time his son was born he was minister for agriculture and a wealthy man. He seems to have had an interest in meteorology, because he purchased meteorological instruments for some local scientific societies. Léon was educated by a tutor who introduced him to meteorology, and the boy developed his meteorological skills working in a small weather station in the south of France. At some time in the 1870s he became secretary of the Société Météorologique de France. The Bureau Central Météorologique (BCM) was formed in 1878 and Teisserenc de Bort obtained a job there soon afterward. He became director of the BCM general meteorology service in 1880. His work involved organizing

The atmosphere consists of distinct layers identified mainly by the way temperature changes with height within them.

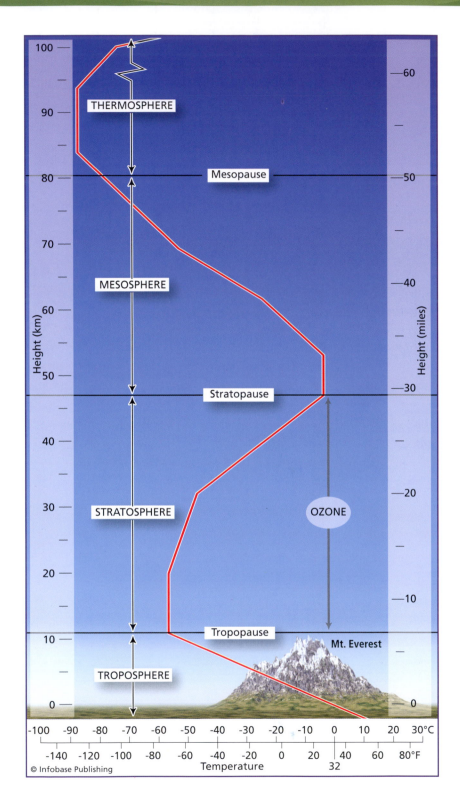

© Infobase Publishing

weather stations in French colonies and on ships and he undertook three expeditions to North Africa, in 1883, 1885, and 1887, to study geology and geomagnetism as well as to supervise the establishment of weather-observing stations. During this period Léon Teisserenc de Bort discovered a link between the distribution of air temperature and pressure, so that high temperature is associated with low pressure and low temperature with high pressure. He was able to demonstrate through this that the location of centers of high and low pressure were associated with weather patterns and that there are a definite number of these centers, which move about average positions. For example, there is a low-pressure center approximately over Iceland (the Iceland low) and a high-pressure center over the Azores (the Azores high), and sometimes the Azores high moves westward, then being known as the Bermuda high. In 1892 Teisserenc de Bort became chief meteorologist at the BCM.

Teisserenc de Bort also participated in an international program on cloud classification, and in 1895 the International Meteorological Committee (precursor of the World Meteorological Organization) planned the International Year of Clouds—a one-year program to measure cloud heights and speeds. This led to the 1896 publication of the first edition of the *International Cloud Atlas.* Léon Teisserenc de Bort was placed in charge of establishing the French station and used his own money to buy a 7.5-acre (3-ha) site at Trappes, near Versailles, where he opened a private meteorological observatory. Scientific work at Trappes began in 1896, measuring clouds, but Teisserenc de Bort realized that the methods he used, involving photographing clouds and measuring their velocity with ground-based instruments, gave access only to the lowest part of what he called "our aerial ocean." That is when he began sampling air at higher levels, first using kites and later balloons.

Léon-Philippe Teisserenc de Bort died at Cannes on January 2, 1913. After his death, his heirs presented the Trappes observatory to the nation, in order that meteorological research might continue there. There is still a meteorological laboratory at Trappes.

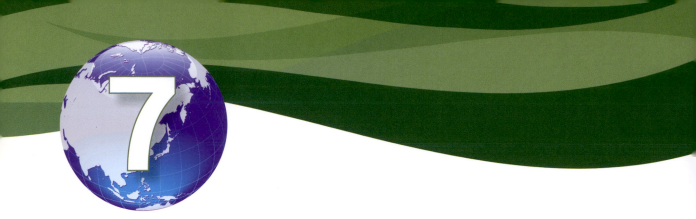

Over Continents
and Oceans

Throughout history until the middle of the 19th century, scientists were able to examine the atmosphere only at a local scale. With the introduction of balloons they could penetrate the upper air, but only the air directly above their heads. They had no way of knowing about atmospheric conditions far away on the other side of a continent or, indeed, on a different continent.

That situation changed with the introduction of the telegraph in the second half of the 19th century (see "Joseph Henry, Samuel Morse, and the Telegraph" on pages 179–185). Meteorologists then began to compile data gathered over a wide area and from those data they were able to construct pictures of weather patterns over an entire continent or ocean.

This chapter tells what happened when scientists considered the implications of this new perspective. It describes the discoveries made by the team of scientists who worked at Bergen, Norway, in the first decades of the 20th century. Their work gave us the terms *air mass* and *front,* and a little later *depression.* The sidebar on pages 152–153 explains what these mean.

It also describes how an attempt to predict the Asian monsoon led to the discovery of a major climatic oscillation associated with a phenomenon known as *El Niño.* This was the first such oscillation to be identified. Climate scientists now know of several and of the ways they affect the weather over large areas. These distant effects are called *teleconnections,* and they are extremely important in long-range weather prediction.

VILHELM BJERKNES AND THE BERGEN SCHOOL

Thanks to the introduction of the telegraph, by the start of the 20th century meteorologists were able to gather data from widely scattered weather stations, and they were developing techniques for forecasting the weather. Scientists recognized that weather systems result from the large-scale motion of the atmosphere, and in 1898 Vilhelm Bjerknes (1862–1951), professor of mechanics and mathematical physics at the University of Stockholm, developed a physical law describing how fluid circulates and forms vortices in the absence of friction. Bjerknes then applied this law to the movement of ocean currents and to the atmosphere. The Carnegie Institution in Washington became interested in his work and supported it, and in 1905 Bjerknes visited the United States. He was appointed a Carnegie research associate with a regular income, and the financial support he received allowed him to employ a long series of assistants—his "Carnegie assistants"—some of whom remained close colleagues and friends for many years. The following illustration shows Vilhelm Bjerknes as he appeared at around this time.

In 1912 Bjerknes became the first professor of geophysics at the University of Leipzig, Germany, where he established a Geophysical Institute that specialized in studying the atmosphere. The team of scientists Bjerknes attracted to the institute came to be known as the Leipzig School. In 1917 Bjerknes was persuaded to return to his native Norway to found and head a Geophysical Institute at the Bergen Museum, as professor of geophysics. The colleagues who gathered there were known as the Bergen School, and the Geophysical Institute is now part of the University of Bergen, specializing in meteorology and oceanography. Bjerknes was 55 years old when he moved to Bergen, but it was there, over the next seven years, that he did his most important work.

Bjerknes and his colleagues established a network of weather-observing stations all over Norway, collecting their data at Bergen. The reports they received allowed them to compile a picture of the weather conditions over Norway at particular times, and

Vilhelm Friman Koren Bjerknes (1862–1951) as a young man. Bjerknes was the Norwegian meteorologist who founded the Bergen Geophysical Institute in 1917. It was there that he did his most important work, establishing the basis of most modern meteorological theory. *(Science Photo Library)*

through studying them they came to realize that over quite large areas the air was at approximately the same temperature, pressure, and humidity throughout. There were extensive blocks of air with distinctive characteristics lying adjacent to other blocks with quite different characteristics. They called these large blocks air masses. Because of their different densities, air in one air mass did not mix readily or rapidly with air in an adjacent air mass. At the time, Europe was at war, and newspapers were full of reports from the front. Air masses reminded the scientists of opposing armies, and the boundaries between them, where the air competed but did not mix, were like the front line. They called these boundaries fronts (see sidebar below).

One front was of particular importance. It separates tropical air to the south from polar air to the north, and it marks where the direct polar and indirect Ferrel cells meet (see sidebar "The Three-Cell Model" on page 128). The Bergen meteorologists called this the polar front and Jacob Bjerknes (1897–1975), Vilhelm's son, developed the theory of the polar front to explain how cyclones form in middle latitudes (see "Jacob Bjerknes and Depressions" on pages 154–157).

The behavior of air obeys the laws of physics and is described mathematically by the gas laws. Vilhelm Bjerknes and his team believed it should be possible to apply the gas laws in order to improve the reliability of weather forecasts. Bjerknes developed a set of seven equations that he believed would allow forecasters to predict the large-scale movement of air. These equations amounted to what he called a graphical calculus, which could be applied to the data describing air across a weather map. Unfortunately the large number of calculations involved and the lack of sufficiently detailed data made the method too slow and unreliable to be practicable, but the approach inspired the English meteorologist Lewis Fry Richardson (1881–1953) to develop it further (see "Lewis Fry Richardson, Forecasting by Numbers" on pages 196–199).

Vilhelm Friman Koren Bjerknes was born on March 14, 1862, in the Norwegian city of Christiania (now called Oslo). His father, Carl Anton Bjerknes (1825–1903) was professor of applied mathematics at the University of Christiania. Carl was not good at performing experiments, and Vilhelm often helped him with his investigations of hydrodynamics. Vilhelm wrote his first scientific paper, "New

Hydrodynamic Investigations," in 1882, when he was 20. Vilhelm entered the University of Kristiania in 1880 (the spelling was changed from Christiania to Kristiania in 1877), obtaining his master's degree in 1888. The following year Vilhelm ended the collaboration with his father and moved to Paris, where he attended lectures on electrodynamics by the French physicist and mathematician Jules-Henri Poincaré (1854–1912). From 1890 until 1892 Vilhelm worked as an assistant to the German physicist Heinrich Hertz (1857–94) in Bonn, Germany, studying electrical resonance. He returned to Norway in 1892 to complete the work he had done in Bonn for his doctoral thesis, receiving his doctorate in 1892. In 1893 Bjerknes obtained a post as a lecturer at the Högskola (School of Engineering) in Stockholm, and in 1895 he became professor of mechanics and applied mathematics at the University of Stockholm. (Norway and Sweden were united as a single country until 1905.) In 1907, after independence, Bjerknes returned to Norway as professor of mechanics and mathematical physics at the University of Kristiania. He moved to Leipzig in 1912 and to Bergen in 1917. Several of his Bergen colleagues became eminent scientists. These included Jacob Bjerknes, Vagn Walfrid Ekman (1874–1954), Tor Bergeron (1891–1977; see "Tor Bergeron and How Raindrops Form" on pages 85–88), and Carl-Gustav Rossby (1898–1957). In 1926 Bjerknes became professor of mechanics and mathematical physics at what since 1925 had been called the University of Oslo, where he remained until his retirement in 1932.

Vilhelm had married Honoria Sophia Bonnevie, a student of natural sciences at Kristiania, in 1893. They had four sons. Honoria was a warm, hospitable person who contributed greatly to the welcoming atmosphere surrounding the family, and Vilhelm was a gifted and popular teacher who also possessed a talent for attracting the most able scientists to work with him and who made sure their work received full recognition. Honoria died in 1928, but Vilhelm remained active. He had received many honors. In 1932 he became president of the Association of Meteorology of the International Union of Geodesy and Geophysics; in 1936 he organized a meeting of the Association in Edinburgh, Scotland; he addressed the Leipzig Geophysical Institute in 1938; and in 1946 he attended the Newton Tercentenary Celebrations in England. Vilhelm Bjerknes died in Oslo on April 9, 1951.

(continues on page 154)

AIR MASSES AND FRONTS

Air that covers a large area of land or sea acquires uniform characteristics. At every height, the air pressure, temperature, and humidity are fairly constant everywhere. A large body of air with approximately uniform physical characteristics is called an air mass. Although each air mass has properties that are similar throughout, those properties differ markedly from the properties of adjacent air masses that acquired their characteristics over a very different surface or in latitudes far to the north or south.

Air masses are classified according to the areas where they form, known as their *source areas*. The first division is between continental and maritime air masses, abbreviated as c and m, respectively. They are then distinguished by latitude, as arctic (A), polar (P), tropical (T), and equatorial (E). The classifications are then combined to produce seven types of air mass. These are, with their abbreviations as follows:

> continental arctic (cA)
> continental polar (cP)
> continental tropical (cT)
> maritime tropical (mT)
> maritime polar (mP)
> maritime arctic (mA)
> maritime equatorial (mE)

It is possible in principle for cT air to mix with mE air to produce continental equatorial air, but this is not included in the list because oceans cover most of the equatorial region and, although continental equatorial air is a possibility, in fact it never occurs.

Additional letters are sometimes used to qualify the basic type descriptions. These indicate whether air is warmer (w) or cooler (k) than the surface beneath it. For example, if mT air crosses a continent in winter it is likely to be warmer than the land surface, so it would be described as mTw air. In summer, when the continental land surface is warmer than the sea surface, mT air crossing the continent would be mTk air. Cool air often brings gusty conditions; warm air brings stable conditions.

The boundary between adjacent air masses is called a front, and as air masses move the fronts also move. There is very little mixing of the air on either side of a front, because the difference in temperature between warm and cool air means they have different densities, so warm air tends to ride over cool air and cool air tends to undercut warm air. A front is defined as warm or cold by the air behind it. If the air behind a moving front is warmer than the air ahead of the front, it is a warm front. If the air behind the front is cooler than the air ahead of the front, it is a cold front. The terms *warm* and *cold* are relative and imply no particular temperature, provided the air temperature differs from that of air ahead of the front.

Waves often develop along fronts, and these can grow into frontal systems with a center of low pressure, called a depression, at the wave crest. Frontal systems often produce clouds and precipitation, because cold air is pushing beneath warm air and forcing it to rise up the cold front, but fronts can also be weak, with little or no clouds. Eventually the warm air behind the warm front is lifted clear of the surface. The cold front overrides the warm front, and the fronts are said to be occluded or to form an occlusion. The fronts then merge and the warm air dissipates. The illus-

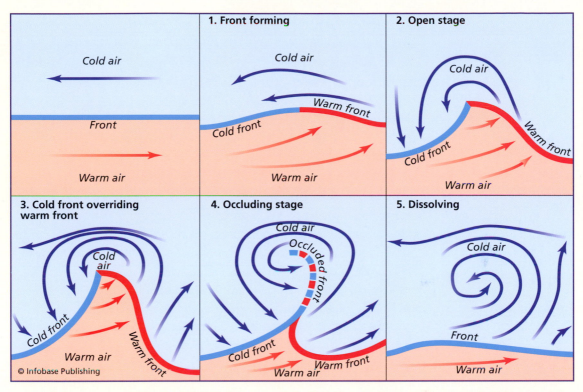

A front between two air masses has warm air on one side and cold air on the other, and the air moves in opposite directions on either side of the front. A wave develops in the front, dividing the front into distinct warm and cold fronts, defined by the air behind them as they move. This is the first of five stages in the development of a frontal system, which ends with the warm air rising clear of the surface and dissipating, and the restoration of the original single front.

tration shows the sequence of events in the life cycle of a frontal system.

Fronts slope. Warm fronts slope at between 0.5° and 1°. Cold fronts slope more steeply, at about 2°. When the first wisps of cirrus cloud indicate the approach of a warm front, the place where the front reaches the surface is about 350–715 miles (570–1.150 km) away. The horizontal distance between the top of a cold front and the place where the front reaches the surface is about 185 miles (300 km).

Fronts are represented on weather maps as lines. A cold front is colored blue or indicated by a black line with small triangles along its leading edge. A warm front is colored red or indicated by a black line with small semicircles along its leading edge. The map shows the position of the fronts on the surface. Despite appearing as lines, however, fronts are really zones 60–120 miles (100–200 km) wide, within which the temperature gradient is steepest, often up to 14.5°F per 100 miles (5°C per 100 km).

(continued from page 151)

JACOB BJERKNES AND DEPRESSIONS

Air flows from all directions toward an area of low atmospheric pressure. As it approaches, the air begins to turn, following a spiral path, until it flows around the low-pressure center parallel to the isobars (see "Christoph Buys Ballot and His Law" on pages 122–125). A general movement of air from all directions toward a central point is called convergence, and its opposite is called *divergence.*

When two adjacent air masses move at different speeds air is forced to rise. This produces a region of low pressure in the lower atmosphere, causing convergence as surrounding air flows inward to fill the low pressure. At high level the rising air flows outward, and this divergence draws more air from below. As air rises, its temperature falls and its water vapor condenses, producing clouds and precipitation. Convergence is associated with cloudy skies—usually with stratiform clouds (see sidebar "Modern Cloud Classification" on page 91)—and often with persistent drizzle, rain, or snow.

The boundary between the colliding air masses came to be called a front, but it was first noticed as a convergence in the wind pattern reported from surface weather stations around Norway. Areas of convergence moved along the Norwegian coast. The scientist at the Bergen Geophysical Institute who realized what was happening was Jacob (later called "Jack") Bjerknes (1897–1975). He saw that cold air was undercutting warmer air, so that warm air was ascending and cold air subsiding. This produced clouds and precipitation in the warm air and stable conditions, with fine weather, in the cold air. A circular flow of air is called a cyclone, from *kuklos,* the Greek word for circle, and in 1919 Jacob Bjerknes published a paper in *Geofysiske Publikasjoner,* "On the Structure of Moving Cyclones," describing what he had discovered. His description of the development and dissolution of a cyclone is still used. He suggested, correctly, that fronts slope, with air rising up the frontal surfaces, according to a mathematical formula devised in 1903 by the Austrian meteorologist Max Margules (1856–1920), and he suggested that the formation of cyclones forms part of the general circulation of the atmosphere, helping transport warm air northward and cool air southward. He also noted that cyclones tend to follow one another, forming groups; these are now called families. Bjerknes was then 20 years old. His paper aroused intense interest among the other scientists at Bergen, and in the years that followed they filled in many of the details of the process. The illustration on page 153 shows the modern idea of the process, but this is based very largely on Jacob Bjerknes's original paper. In 1922 Jacob Bjerknes moved to Zürich as a consultant to the Swiss Meteorological Institute. While there he was able to use data from weather stations in the Alps to measure the slope of fronts. In the 1930s, using data from radiosondes, he showed how the cold front of one cyclone turns into the warm front of the next cyclone in the sequence.

Bjerknes was describing midlatitude cyclones, but there are also tropical cyclones, known as hurricanes in the North Atlantic and Caribbean, typhoons in the Pacific and China Seas, and cyclones in the Indian Ocean. To avoid confusion, midlatitude cyclones are often called depressions. The difference in the names reflects different points of view. "Cyclone" refers to the way air circulates; "depression" refers to the low pressure at the center.

The prevailing winds in middle latitudes blow from the west, and in the 1930s meteorologists of the Bergen School, including Jacob Bjerknes, had calculated that the strength of these winds must increase with altitude, reaching a maximum near the tropopause. During World War II American aircrews flying over the Pacific and German aircrews flying over the Mediterranean discovered the dramatic truth of these calculations. They found that sometimes—but not always—their east-west journey times were longer than they had predicted and west-east flights were shorter. They concluded that there are winding ribbons of fast-moving air at high altitudes blowing in a general westerly direction. The aircrews called these ribbons of wind jet streams and learned to measure the rate at which their ground speed changed as they approached them. After the war Jacob Bjerknes undertook a major international research project into the jet streams and their relationship to the general circulation of the atmosphere and to the formation and movement of frontal systems.

In later years Bjerknes turned to the interaction between the atmosphere and the oceans and he came to recognize that variations in climate on the scale of decades were generated within this ocean-atmosphere system. He studied the way variations in the midlatitude westerly winds affect the Gulf Stream and the temperature of the North Atlantic. He also studied the Pacific Ocean and especially the periodic change in pressure distribution, winds, and ocean currents known as El Niño (see "Gilbert Walker, Oscillations, and El Niño" on pages 157–161).

Jacob Alle Bonnevie Bjerknes was born in Stockholm on November 2, 1897. His father was Vilhelm Bjerknes (see "Vilhelm Bjerknes and the Bergen School" on pages 149–152) and his mother was Honoria Sophia Bonnevie. He was named after Honoria's father, Jacob Alle Bonnevie, who was Norwegian minister for education. Jacob's aunt, Kristine Bonnevie, was a zoologist and Norway's first female professor. It was a highly intellectual family on both sides. Jacob's education began in Stockholm, but in 1907, when he was nine, the family moved to Kristiania. When his parents moved to Leipzig in 1913 Jacob remained behind to complete his education at junior college and to enroll at the University of Kristiania, but in 1916 he left the university and went to join the family at Leipzig. There he took over research into wind convergence that a German student, Herbert Petzold, had started as a doctoral project. World War I had started in 1914. Petzold had been drafted into the army and was killed at Verdun. As the war continued conditions in Leipzig became increasingly difficult, and in 1917 the family moved to Bergen. Jacob became a full member of the Bergen School, known affectionately as young Bjerknes.

In 1920 Jacob was made head of weather forecasting for western Norway. He married in 1928. His wife, Hedvig Borthen, was the daughter of an ophthalmologist. Jacob left his post with the weather service in 1931 to take up a professorship in meteorology that had been created for him at the Bergen Museum.

Accompanied by his family, in July 1939 Jacob traveled to the United States for an eight-month lecture tour, but on September 1 German forces invaded Poland and on September 3 Britain and France declared war on Germany. World War II had commenced, and in April 1940 Germany invaded and occupied Denmark and Norway. The Bjerknes family remained in the United States and in 1946 they became American citizens.

Jack Bjerknes (his American name) was asked to establish a school based at the University of California for training meteorological officers for the U.S. Army Air Corps. Jack chose the Los Angeles campus because of its proximity to the Scripps Oceanographic Institution. In 1940 he became professor of meteorology and head of the meteorology section in the department of physics. In 1945 Bjerknes became head of a newly formed department of meteorology, which soon became a major center for meteorological research and teaching. He died in Los Angeles on July 7, 1975.

GILBERT WALKER, OSCILLATIONS, AND EL NIÑO

Not surprisingly, most of the major discoveries in the atmospheric sciences are made by atmospheric scientists—physicists, chemists, meteorologists, or climatologists. But there is a notable exception. Gilbert Walker (1868–1958) was a brilliant mathematician with a special interest in spinning tops and projectiles. He was also an authority on the mathematics of various ball games and on the aerodynamics of boomerangs, which he could throw with great skill. It was Gilbert Walker who discovered the atmospheric processes that produce El Niño events. Today most people have heard of El Niño. El Niño events occur at intervals of one to seven years, when the trade winds weaken and occasionally cease or even reverse direction over the tropical South Pacific Ocean.

The trade winds drive the South Equatorial Current, which flows parallel to the equator from east to west, and when the winds weaken or reverse the current also weakens and may even reverse direction. The South Equatorial Current carries warm surface water across the ocean in the direction of Indonesia, where it accumulates to form a deep pool of warm water called the warm pool. On the opposite side of the ocean, the layer of warm water near the South American coast is much shallower and the Peru Current, which flows northward parallel to the coast, has many upwellings that bring to the surface cold water that the current has carried all the way from Antarctica. Evaporation from the warm pool gives Indonesia a very wet climate, but the cold water off the South American coast exacerbates the extremely arid conditions prevailing there.

During an El Niño this pattern weakens or reverses. Warm water accumulates off South America, suppressing the upwellings of cold water and producing heavy rain. Around Indonesia, where the surface water is cooler than normal, there are droughts. The climatic effects are also felt much more widely. The change becomes evident in late December, and the rain it brings to the parched lands of Peru and Chile led Christian missionaries to call it El Niño, the boy child, because they saw it as a Christmas gift that meant food would be abundant in the following year.

In 1903 Gilbert Walker became a special assistant to Sir John Eliot (1839–1908), whose official title was Meteorological Reporter to the Government of India and Director-General of Indian Observatories. On December 31, 1903, Sir John retired, and on January 1, 1904, Walker became head of the India Meteorological Department, based at Shimla (then called Simla).

The Indian government asked Walker to search for some way to predict the Indian monsoon, and almost from his first day that is what he set out to do, submitting papers on the monsoons and comparisons of predicted and actual rainfall that the Meteorological Department published in 1906. It was a vital task, because Indian farmers depend on the monsoon rains. If these are late the harvest may be poor or fail entirely, and if they are excessive the land may flood, also destroying the crops. Life turns on the annual rains, and they are notoriously unreliable. Modern communications and the ability to stockpile food reserves mitigate the effects, but historically a failure of the monsoon has meant famine, and when Walker took over at the Meteorological Department two severe famines had occurred in recent times. More than 5 million people died in India during the famine of 1876–1877 and the 1899 famine cost 4.5 million lives. The greatest of all the monsoon famines was in Bengal in 1770. It killed up to one-third of the population. If the timing and strength of the wet summer monsoon could be predicted it might be possible to distribute emergency supplies to regional centers in advance and to advise farmers on dates for sowing.

Walker found he could not predict the monsoon simply by applying mathematics to the problem. Instead, he collected data on weather events throughout southern Asia and on the monsoon and used statistical techniques to seek patterns. These slowly emerged. He noted that it is the monsoon rainfall of Ethiopia that determines the timing of the annual Nile flood on which Egyptian farmers depend, and that the winds bringing the rain to Ethiopia cross the Indian Ocean alongside the winds that bring the Indian monsoon.

The emerging patterns revealed that the easterly trade winds are balanced by winds at high level, near the tropopause, that flow in the opposite direction, from west to east. These form the upper part of a series of cells, now called *Walker cells,* that are superimposed on the Hadley cells (see "Edmond Halley, George Hadley, and the Trade Winds" on pages 115–118) and feed the trade winds. The following

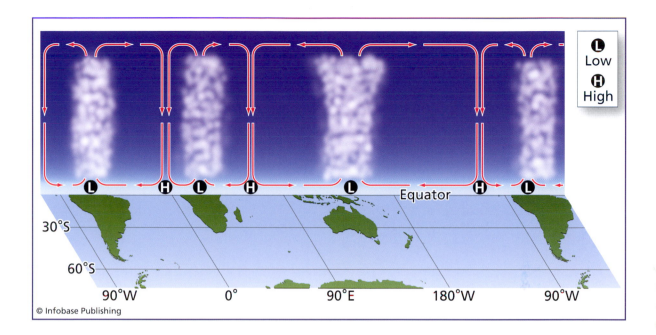

L Low
H High

Equator

30°S

60°S

90°W   0°   90°E   180°W   90°W

© Infobase Publishing

illustration shows this pattern of Walker cells, forming the *Walker circulation.*

Walker also discovered another pattern. He found that in some years the surface atmospheric pressure is lower than usual around Indonesia and the eastern side of the Indian Ocean, and higher than usual over Rapa Nui (formerly called Easter Island) in the South Pacific Ocean. Low pressure is associated with convergence, rising air, and heavy precipitation, high pressure with subsiding air, clear skies, and dry weather. This distribution of pressure strengthens the trade winds over the South Pacific, because air is flowing outward from the high-pressure region in the east and into the low-pressure region in the west. In other years the pattern was reversed, with higher than usual pressure over Indonesia and lower pressure over Rapa Nui. Walker found that the distribution pattern changes constantly, reaching maxima at intervals of two to seven years. A periodic change of this kind is an oscillation, and Walker called this one the *Southern Oscillation* because of where it occurs. It is the Southern Oscillation that produces El Niño events and their opposite, *La Niña* events. The complete cycle is now known as an *ENSO* (El Niño–Southern Oscillation) event and it can be predicted, partly through monitoring changes in pressure over Tahiti and Darwin, Australia,

The circulation pattern discovered by Gilbert Walker (1868–1958) in 1923 carries air at high level in the opposite direction to the surface trade winds in a series of cells imposed on the Hadley cells.

to detect the Southern Oscillation. The Southern Oscillation was the first climate oscillation to be discovered. Since then several more have been identified, including the North Atlantic Oscillation, which Walker discovered and named, and all of them have widespread climatic effects.

Gilbert Walker was born on June 14, 1868, at Rochdale, Lancashire, an industrial town in northwestern England. He was the eldest of eight children. His father was a civil engineer, who, soon after Gilbert was born, obtained the post of chief engineer to the town of Croydon, near London, and the family moved south. Gilbert entered a local school in September 1876 and in 1881 he won a scholarship to St. Paul's School, in London, where he excelled in mathematics. In 1884 he entered Trinity College, University of Cambridge, graduating in mathematics in 1885. During the late 1880s Walker visited Australia, where his interest in boomerangs began.

Walker won many prizes at Cambridge, was made a fellow of Trinity College in 1891, and received his master's degree in 1893. The intense hard work damaged his health, however, and he had to spend the winters of 1890, 1891, and 1892 in Switzerland, where he became an expert skater and keen mountaineer. Walker was made a lecturer in mathematics in 1895, but resigned both his lectureship and fellowship in 1903 in order to take up the post in India. He married May Constance Carter in 1908, and they had one son and one daughter.

In December 1924 Gilbert Walker retired from his Indian post and became professor of meteorology at Imperial College of Science and Technology, London. While there he continued his meteorological research, but also experimented more widely in physics, concentrating on convection in unstable fluids and the formation of clouds. He published several papers on bird flight and the effects weather had on it. He was a keen musician and played the flute, and he enjoyed sketching and painting. Walker retired from Imperial College in 1934 and moved to Cambridge. In 1950 he moved again, in the following years living in several places, mainly in Sussex and Surrey, south of London.

Gilbert Walker was elected a fellow of the Royal Society in 1904 and received the degree of doctor of science in the same year from the University of Cambridge. He was elected a fellow of the Royal

Meteorological Society in 1905 and was its president in 1926 and 1927. He was a fellow of the Royal Astronomical Society and an honorary fellow of Imperial College. He was knighted in 1924. His wife died in 1955, and Sir Gilbert died on November 4, 1958 at Coulsdon, Surrey.

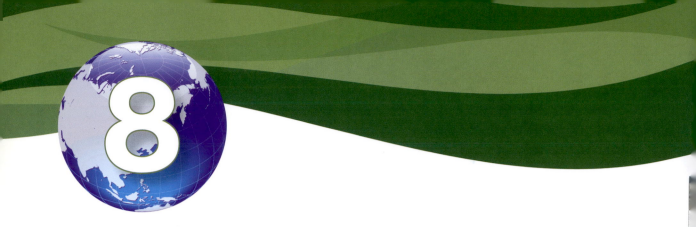

Classifying Climates

It has always been obvious to travelers that climates vary from place to place. In Europe, for example, places near the Atlantic coast have milder winters, cooler summers, and more rainfall than places farther east. Meteorologists now know this is because midlatitude weather systems move from west to east, bringing moist maritime air to coastal regions of Europe, but the air loses its moisture as it drifts inland. This effect is very marked where a mountain range runs north to south, at right angles to the approaching weather. Places on the western side of the Rocky Mountains in the northern United States and British Columbia have a moister, more temperate climate than those to the east of the mountains. In the Tropics, however, the effect is reversed because there the prevailing winds bring weather systems from the east, so air must cross the continent and then loses such moisture as it still retains as it crosses the mountains. The Atacama and Namib Deserts are located on the western coasts of South America and Africa, respectively, but in the Tropics.

Throughout history philosophers and scientists have tried to classify climates. Today there are many systems of classification to choose from and each has its advantages and disadvantages. They fall into two broad categories: generic and genetic. Generic classifications are based mainly on temperature and precipitation and their effects on plant growth. Genetic classifications are based on those features of the atmospheric circulation that produce each type of climate.

This chapter describes the origin of climate classification, in the ancient world. The earliest scheme might be described as a genetic classification, and the chapter also mentions one example of a more modern genetic scheme. The chapter than describes the history of the two systems most widely used today. These were devised by the German meteorologist W. P. Köppen (1846–1940) and the American meteorologist C. W. Thornthwaite (1899–1963). Both are generic classifications.

TORRID, TEMPERATE, AND FRIGID CLIMATES

The most obvious way to categorize climates is by latitude. Travel away from the equator and the temperature falls. Travel toward the equator and the weather becomes warmer. The earliest climate classifications were based on latitude. In modern terms they might be described as genetic classifications, because latitude determines the amount of solar radiation the surface receives and this, in turn, determines the temperatures and rate of evaporation.

Aristotle (384–322 B.C.E.) formalized a system that was already familiar to many people. In Book II, Part 5 of his *Meteorologica*, written in about 350 B.C.E., he states (in the translation by E. W. Webster):

There are two inhabitable sections of the Earth: one near our upper, or northern pole, the other near the other or southern pole; and their shape is like that of a tambourine. If you draw lines from the center of the Earth they cut out a drum-shaped figure. The lines form two cones; the base of the one is the Tropic, of the other the ever visible circle, their vertex is at the center of the Earth. Two other cones towards the south pole give corresponding segments of the Earth. These sections are habitable. Beyond the Tropics no one can live: for there the shade would not fall to the north, whereas the Earth is known to be uninhabitable before the Sun is in the zenith or the shade is thrown to the south: and the regions below the Bear are uninhabitable because of the cold.

Aristotle shows from this that climatic belts extend all the way around the Earth. "If we reflect we see that the inhabited region is limited in breadth, while the climate admits of its extending all round the Earth." Consequently, he says, there is nothing to prevent people from traveling all the way round the world, except that oceans bar their progress. On the other hand it is impossible to travel very far to the north or south, because once outside the habitable belt conditions quickly become intolerable.

The Greeks had not actually visited the Southern Hemisphere and Aristotle was telling them that it would be impossible for anyone living in the Northern Hemisphere to do so, because it would mean crossing a wide belt where the heat would be intolerable. Nevertheless, he could reason that the Southern Hemisphere must resemble the Northern Hemisphere because the Greeks had a geometric view of the universe that required the Earth to be symmetrical. He proposes that just as winds blowing from the north bring air from the North Pole, there must be winds in the Southern Hemisphere that carry air from the South Pole, but these will be southerly winds. In the Northern Hemisphere southerly winds are warm because "the south wind clearly blows from the torrid region."

Having determined that the habitable regions of the world are confined to a belt around each hemisphere, the next step was to define boundaries. This Aristotle was able to do by geometry. He based his calculations on the positions of the rising and setting Sun at the *equinoxes* and *solstices.* From these he was able to calculate the position of the equator and the two Tropics, of Cancer in the Northern Hemisphere and Capricorn in the Southern Hemisphere, and of the Arctic and Antarctic Circles. His climate classification was now complete. Two Frigid Zones extended from the North and South Poles as far as the Arctic and Antarctic Circles. These were uninhabitable. Two Temperate Zones extended from the Arctic and Antarctic Circles to the Tropics. These were the habitable zones. Two Torrid Zones extended from the equator to the Tropics on either side. The following illustration shows Aristotle's climate system.

Aristotle knew the significance of the equinoxes and solstices. He knew that at the equinoxes, now on March 20–21 and September 22–23, the Sun is above the horizon for 12 hours and below it for 12 hours, and that where he lived, in Greece, the solstices, now on June 21–22 and December 22–23, are midsummer day, with the longest days and shortest nights, and midwinter day, with the shortest days and longest nights, respectively. He also knew that the equinoxes define the Tropics. However, he would not have known why this is so

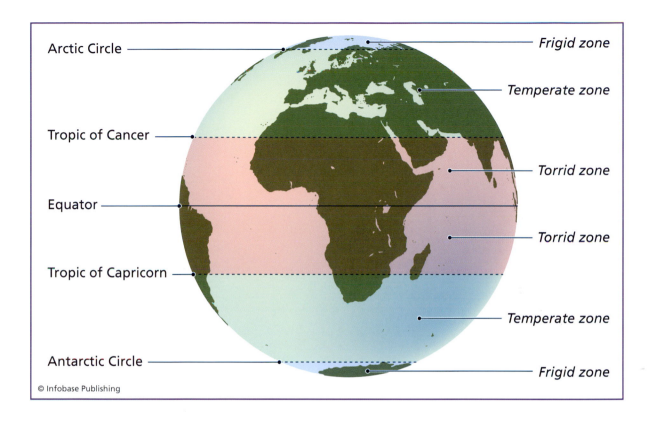

Arctic Circle — Frigid zone

Temperate zone

Tropic of Cancer — Torrid zone

Equator

Torrid zone

Tropic of Capricorn

Temperate zone

Antarctic Circle — Frigid zone

© Infobase Publishing

or what the connection is between the latitudes of the Tropics and of the Arctic and Antarctic Circles. The reason is that Earth is tilted on its rotational axis by 23°30′ (23.5°). Consequently, as Earth orbits the Sun, first one hemisphere and then the other is tilted toward the Sun and enjoys summer. The exposure to the Sun reaches its minimum and maximum at the solstices, and the equinoxes occur at the halfway points between the solstices. The illustration that follows shows how the Earth's tilt produces the seasons, with the solstices and equinoxes.

The tilt also defines the tropics of Cancer in the Northern Hemisphere and Capricorn in the Southern Hemisphere. Because of the 23°30′ axial tilt, at the solstices the noonday Sun is at an angle of 23°30′ to the vertical from any point on the equator and it is directly overhead at latitude 23°30′, in one or other hemisphere. The Tropics, therefore, are lines around the Earth marking the farthest distance from the equator (the highest latitude) at which the Sun is directly overhead at noon on one day in the year. The equinoxes are the two

Aristotle (384–322 B.C.E.) classified the world's climates by latitude, a method that was used widely throughout the ancient world and which he formalized.

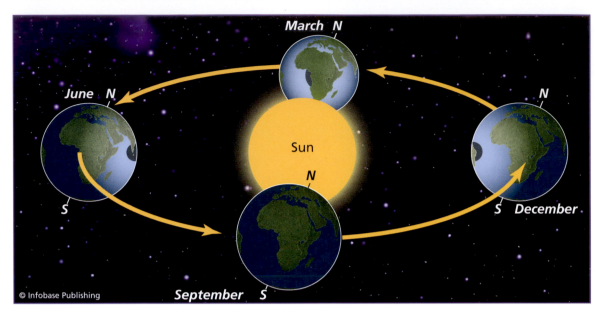

The Earth's axis is tilted with respect to the plane of its solar orbit. Consequently, in June the Northern Hemisphere faces the Sun, and in December the Southern Hemisphere faces the Sun. At the March and September equinoxes, both hemispheres are exposed equally.

days each year when at noon the Sun is directly overhead at the equator. The illustration shows why this is so. The tilt also defines the Arctic and Antarctic Circles, which mark the farthest distances from each pole at which there is at least one day each year when the Sun does not rise above the horizon and one day when it does not sink below the horizon. These boundaries are located at latitudes 66°30′ N and S (90 - 23.5 = 66.5).

Aristotle's was the first genetic classification of climate, but it was not the last. In 1950 the German meteorologist Hermann Flohn (1912–97) proposed another. Flohn divided climates into eight groups, based mainly on the global wind belts in which they occur and the amount and type of precipitation each receives. For example, he defined a subtropical dry zone, dominated either by the trade winds or by the subtropical region of semipermanent high pressure, where conditions are perpetually arid, and an extratropical westerly zone, where there is precipitation throughout the year. Flohn was

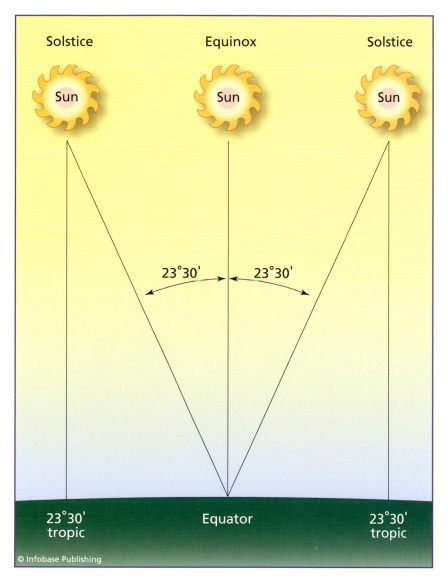

The Earth's axis is tilted by 23°30′ with respect to the plane of its solar orbit. Consequently, there are two days each year, called the equinoxes, when at noon the Sun is directly overhead at the equator. There are also two days each year when at noon the Sun is at an angle of 23°30′ to a vertical line drawn at the equator. On these days, therefore, the Sun at noon is directly overhead at latitude 23°30′ N or S. These days are the solstices, and these latitudes define the Tropics.

professor of meteorology at the University of Bonn and one of the most eminent climate scientists of his generation. His climate classification is considered one of the best genetic schemes. He was also one of the first climate scientists to express concern, in a paper he wrote in 1941, about the possibility that burning fossil fuels might be altering the global climate.

WLADIMIR KÖPPEN AND HIS CLASSIFICATION

Generic classifications categorize climates by their temperature ranges and by the type and distribution of precipitation, usually according to the kind of vegetation each climate supports. The most widely used classification of this type first appeared in 1884, when its author, the German meteorologist Wladimir Peter Köppen (1846–1940) published a map of climate types. Köppen went on to develop this into a full classification, covering every type of climate, which he published in 1900. He continued to work on it, publishing a revised version in 1936, and he was still seeking to improve it, by then in collaboration with the German climate scientist Rudolf Geiger (1894–1981), until his death.

Köppen began by dividing the world's climates into five principal types designated by capital letters: equatorial (A); arid (B); warm temperate (C); snow (D); and polar (E). He then qualified these by adding letters for precipitation and temperature. In the latest version of his scheme, the precipitation categories are: desert (W); steppe (S); fully humid (f); summer dry (s); winter dry (w); and monsoonal (m). The temperature categories are: hot arid (h); cold arid (k); hot summer (a); warm summer (b); cool summer (c); extremely continental (d); polar frost (F); and polar tundra (T). With this scheme, every climate in the world can be broadly described by a maximum of three letters. New York, for example, has a Cfb climate: warm temperate, fully humid, warm summer. The Sahara Desert is BWh: arid, desert, hot arid. The map shows the Köppen-Geiger classification applied to the world.

The Köppen categories relate to vegetation. For example, trees will not grow where the average summer temperature is lower than 50°F (10°C). Category C is based on winter temperatures between 26.6°F (-3°C) and 64.4°F (18°C) and summer temperatures higher than 50°F (10°C). Category D climates also have summer temperatures higher than 50°F (10°C), but winter temperatures are lower than 26.6°F (-3°C).

Köppen was not the first scientist to notice that particular regions of the world support characteristic types of vegetation, and he had the work of earlier biogeographers on which to build. In 1855 the Swiss botanist Alphonse-Louis-Pierre-Pyramus de Candolle (1806–93) published *Géographie botanique raisonée* (Botanical geography

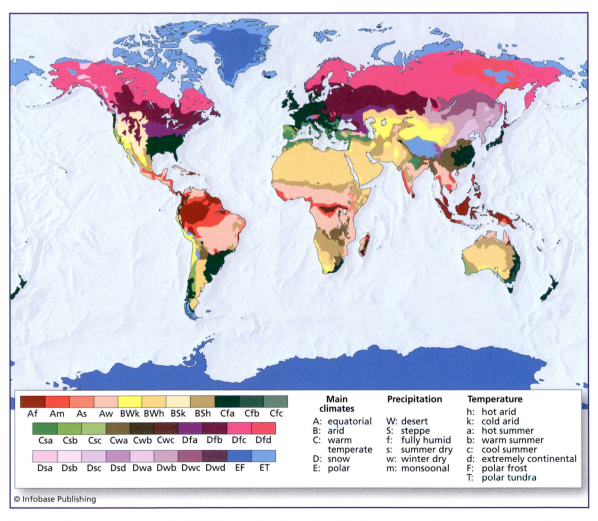

| | | | | | | | | | | | | Main climates | Precipitation | Temperature |
|Af|Am|As|Aw|BWk|BWh|BSk|BSh|Cfa|Cfb|Cfc| | | | |

A: equatorial
B: arid
C: warm temperate
D: snow
E: polar

W: desert
S: steppe
f: fully humid
s: summer dry
w: winter dry
m: monsoonal

h: hot arid
k: cold arid
a: hot summer
b: warm summer
c: cool summer
d: extremely continental
F: polar frost
T: polar tundra

© Infobase Publishing

The map shows the current (April 2006) classification of world climates according to the scheme that Wladimir Peter Köppen (1846–1940) developed, in later years in collaboration with Rudolf Geiger (1894–1981).

classified) and in 1880 he published *La Phytographie* (Phytography). Phytography is the scientific study of the geographic distribution of plants. De Candolle was professor of natural history at the University of Geneva and probably the most eminent botanist of his day. Köppen used de Candolle's map of plant distribution as the basis for his own map of climate types.

Wladimir Peter Köppen was born on September 25, 1846, in St. Petersburg, Russia, to German parents. His grandfather, a physician, had moved to Russia at the invitation of the empress Catherine II (Catherine the Great; 1729–96) as one of a number of foreigners brought in to improve sanitation services in the provinces. Later he became a personal physician to the czar. Wladimir's father, Peter von Köppen (1793–1864), was a geographer and authority on Russian cultures. Czar Alexander II (1818–81) made him an academician, which was the highest academic honor Russia could bestow, and granted him an estate, called Karabakh, in the Crimea.

Wladimir attended a secondary school at Simferopol, near Karabakh, from 1858 until 1864, when he enrolled at the University of St. Petersburg to study botany. In 1867 he moved to the University of Heidelberg and in 1870 to the University of Leipzig, both in Germany. His student dissertation, which he defended at Leipzig in 1870, was on the way temperature affects plant growth. During the Franco-Prussian War of 1870–1871 Köppen served in the Prussian ambulance corps.

After graduating, Köppen was employed by the Russian meteorological service in 1872 and 1873. In 1875 he returned to Hamburg, Germany, to take up a post as head of the division of marine meteorology at the Deutsche Seewarte (German naval observatory). His task there was to establish a weather forecasting service for northwestern Germany and the adjacent sea areas. His primary interest was in research, however, and in 1879, with the service operational, he was given the newly created title of meteorologist to the Seewarte, a position that allowed him more time for research. He made a systematic study of the climate over the oceans and used kites and balloons to investigate the upper atmosphere. He also developed his classification scheme, which he mapped onto an imaginary continent that he called *Köppen'sche Rübe* (Köppen's beet).

Köppen was also interested in past climates, and in 1924 he collaborated with his son-in-law, Alfred Lothar Wegener (1880–1930), in writing *Die Klimate der Geologischen Vorzeit* (The climates of the geological past). Köppen also wrote *Grundrisse der Klimakunde* (Outline of climate science), published in 1931. In 1927 he began a collaboration with Rudolf Geiger to produce a five-volume *Handbuch der Klimatologie* (Handbook of climatology). This was never completed, but parts of it were published.

In 1919 Köppen retired from the Seewarte and in 1924 he went to live in Graz, Austria. He spent his remaining years writing his books. Wladimir Köppen and his wife had five children and they also took into their home a nephew and niece whose parents had died. Köppen loved children and while he lived in Hamburg often spent time working at a boys' home that he had founded. He was also active in several reform movements, seeking changes in schools and improved nutrition for underprivileged children. He was an enthusiastic advocate of Esperanto, the artificial language he believed could contribute to world peace if it were widely adopted. He spoke and wrote it fluently and translated some of his own publications into it. A modest man, he never used "von" before his surname, although he was entitled to do so. Wladimir Köppen died at Graz on June 22, 1940.

C. W. THORNTHWAITE AND HIS CLASSIFICATION

In 1931 the American climatologist C. W. Thornthwaite (1899–1963) published a generic classification scheme that would rival the Köppen classification. In its original version it applied only to North America, but in the years that followed Thornthwaite expanded it to cover the whole world. The final version of the scheme appeared in 1948, in "An approach toward a rational classification of climate," a paper by Thornthwaite that was published in *Geographic Review* (vol. 38, pages 55–94). Many scientists consider the Thornthwaite scheme to be superior to Köppen's classification. It is more rational, and its categories are based on mathematics applied to temperature and precipitation data, rather than the raw data that Köppen used. This makes the Thornthwaite scheme very useful to land-use planners and agricultural economists. The scheme's disadvantages are that the mathematics make it fairly complicated to use, and its climate categories have never been plotted on maps, so there is no way to tell by a glance at a map what kind of climate a particular place has.

Thornthwaite recognized that what matters to farmers is not the amount of precipitation or warmth a place receives, but the extent to which plants are able to take advantage of the conditions. The purpose of the mathematics he introduced was to calculate precipitation efficiency and thermal efficiency. Precipitation efficiency is the annual sum of the monthly ratios of precipitation to evaporation. The calculation takes account of temperature, because temperature determines the rate of evaporation, and the result is presented as a precipitation-efficiency index, which indicates the amount of water available to plants through the year. In his 1948 revision Thornthwaite introduced a moisture index. This shows the month-by-month surplus or deficit of water present in the soil. He also developed a humidity index, indicating the extent to which the amount of moisture available to plants exceeded the amount needed for healthy growth. Thermal efficiency also links temperature to evaporation.

Thornthwaite's most important innovation, which did not appear until the 1948 revision, was his concept of potential evapotranspiration, which is the amount of water that would be lost from the soil by evaporation and plant transpiration if the supply of water were unlimited. This can be measured as the amount of water that evaporates from an open container. Provided the average monthly temperatures are known, the potential evapotranspiration can be read from tables.

The Thornthwaite scheme then categorizes climate types according to the moisture index and potential evapotranspiration. It divides them into moisture provinces and temperature provinces. There are nine moisture provinces, identified by capital letters, some with subscripted numbers, relating to the moisture index value. For example, a B_3 climate is described as humid and has a moisture index of 40–60. Climate D is semiarid, with a moisture index between -60 and -40.

There are nine temperature provinces, also identified by letters with a prime (') to distinguish them from the moisture provinces, and described as frost (E'), tundra (D'), microthermal (C'_1 and C'_2), mesothermal (B'_1, B'_2, B'_3, and B'_4), and megathermal (A'). Each of these categories corresponds to a value for potential evapotranspiration. The temperature provinces are qualified further as moist or dry, indicated by lower-case letters. Moist climates are measured against an aridity index and dry climates against a humidity index.

The classification provides a wealth of detailed information that is directly applicable to agriculture and other forms of land use, but it is rather complicated to interpret, and although the calculations involved in preparing the indices are not mathematically demanding, there are many of them and the indices depend crucially on the availability of good data. Consequently, the classification most students encounter is the much simpler one devised by Wladimir Köppen. The Thornthwaite classification is widely used by professional geographers and planners, however.

Charles Warren Thornthwaite was born on March 7, 1899, near Pinconning, in Bay County, Michigan. He attended Central Michigan Normal School (now Central Michigan University) and the University of Michigan, graduating in 1922 and becoming a teacher in a high school in Owosso, Michigan. He received his doctorate in 1929 from the University of California, Berkeley. From 1926 until 1930 he worked as a geographer with the Kentucky Geological Survey and he held a faculty position as assistant professor of geography at the University of Oklahoma from 1927 until 1934. From 1935 until 1946 he headed the division of climatic and physiographic research of the U.S. Soil Conservation Service.

In 1947 Thornthwaite was appointed professor of climatology at the Johns Hopkins University. While there he established a laboratory of climatology at Seabrook Farms and later at Centerton, New Jersey. In 1954 he transferred the laboratory to the Drexel Institute of Technology, Philadelphia. The laboratory is now run on behalf of C. W. Thornthwaite Associates, which Thornthwaite formed in 1952 to provide consultancy services and manufacture meteorological instruments.

Thornthwaite received many honors and held several public offices. He was president of the meteorological section of the American Geophysical Union from 1941 until 1944. In 1951 he became president of the Commission on Climatology of the World Meteorological Organization, holding this post until his death. In 1961 he was elected honorary president of the Association of American Geographers. Thornthwaite died from cancer after a prolonged illness on June 11, 1963.

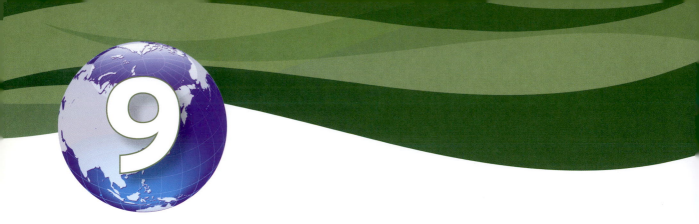

Weather Reports, Maps, and Forecasts

Scientists study the weather primarily in order to be able to forecast it. Although the composition and structure of the atmosphere, and the laws governing the movement of air and the formation of cloud and precipitation are fascinating topics of study, forecasting has always been the underlying aim of atmospheric research. Prior to the 19th century, however, there was only one method for foretelling the weather: by historical analogy. People observed that when certain cloud patterns were seen, or when the wind blew from a certain direction, a particular type of weather would follow. Much weather lore (see sidebar "Weather Lore" on page 9) was based on this type of analogy. A red sky at night very often heralds a fine day to follow. Wisps of cirrus—"mares' tails"—often mean the wind will strengthen. When the air feels hot and oppressive, thunderstorms are more likely. Observations of this kind have rational explanations, but they existed as weather predictors long before meteorologists discovered that dust in approaching dry air produces red sunsets, that cirrus forms at the top of active weather fronts, and that air feels oppressive because both the temperature and relative humidity are high. That there are now explanations for the old sayings does not alter the fact that they have survived because historical analogy works; when the sky looks like this or the air feels like that, experience tells us that this is the weather to expect.

Wind directions are the best predictors of all. In temperate latitudes of the Northern Hemisphere, a wind from the north may carry air from the Arctic. It is a cold wind. A wind from the ocean will be mild and wet, perhaps stormy. A wind blowing from the center of a continent will bring dry air that is hot in summer but bitterly cold in winter.

This chapter describes the development of weather forecasting, and it begins with the world's oldest weather forecasting device, the Tower of the Winds. As its name suggests, it forecast the weather on the basis of wind direction. People knew what weather to associate with each wind direction, and the Tower reminded them, but it could not predict when the wind direction might change.

It is one thing to recognize signs of imminent weather, but quite another to predict the appearance of those signs at a time when the sky looks quite different or the wind shows no indication of changing direction. If meteorologists are to make true forecasts of the weather, they need information gathered from a wide area in order to understand the state of the atmosphere and identify the weather systems it contains and how these are moving. Weather systems are extremely large. The map below shows the weather systems over North America

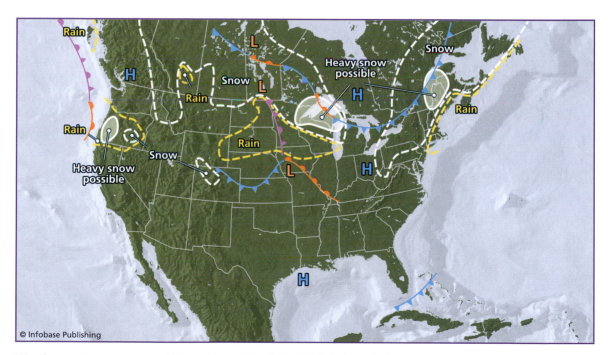

Weather systems over North America on March 12, 2008. A frontal depression is centered over Iowa. Its warm front extends as far as Kentucky, its cold front as far as Colorado, and its occluded section into Manitoba. Another system farther east extends from Alberta, across the Great Lakes, and into Maine.

on March 12, 2008. A frontal depression is centered over Iowa, with a warm front extending to Kentucky, a cold front to Colorado, and occluded fronts extending northward into Manitoba. Another system to the north and east extends from Alberta, across the Great Lakes, and to Maine. It was not until the 19th century that meteorologists could begin to gather the data they needed to produce maps of this kind.

The change came with the invention of the telegraph. This chapter tells how that came about and how almost at once the telegraph was used to gather meteorological data. The first weather map was shown to the public in 1851, and daily weather reports first appeared in 1869. The first forecast to appear in a newspaper was written by Robert FitzRoy, formerly captain of HMS *Beagle,* the ship on which Charles Darwin (1809–82) sailed to South America.

Weather forecasting remained difficult and unreliable, however. The chapter ends with an account of the attempt to prepare forecasts mathematically long before the days of the electronic computer. It then describes the discovery of the inherent problem that makes long-range forecasting impossible.

THE TOWER OF THE WINDS

People have always wanted to know what the weather would be like tomorrow, next week, or next month. For thousands of years the best they could do was to trust to old sayings, most of which were unreliable, or use their own judgment and experience. Then, in the first century B.C.E., someone decided it was possible to do better.

Andronicus was born in about 100 B.C.E. in Cyrrhus, Syria. He was an astronomer, engineer, and architect who made astronomical instruments and celestial spheres showing the positions of the planets and stars. He also built a white marble sundial for the sanctuary of Poseidon, god of the sea, and his wife Amphitrite on the island of Tiros. This sundial became famous throughout the region, and Andronicus was invited to Athens. The Greeks called him Andronikus Kyrrhestes, or Andronicus of Cyrrhus. Little more is known about Andronicus, and there is not even a record of when he died. What is known is that while he was in Athens Andronicus designed and supervised the construction of a horologion, which is a clock in the form of a building. Several contemporary writers described the horologion. One of them, the Roman writer, architect, and engineer Vitruvius (Marcus Vitruvius Pollio, ca. 70–25 B.C.E.), called it the Tower of the Winds.

As the following illustration shows, much of the tower still stands. Now a tourist attraction, originally it must have seemed a marvel. It is located in the Roman agora—marketplace—where everyone could see it. Built from marble, it is 40 feet (12 m) high, and each of its eight sides is 10 feet (3.2 m) long. It once had a conical roof on top of which there was a bronze figure of Triton, who was the son of Poseidon and Amphitrite, holding a rod. Triton has the upper body of a man and the lower body of a fish. His statue was mounted in such a way as to turn with the wind so that his rod pointed in the direction from which the wind was blowing. This was the world's first weather (wind) vane, and every wind vane since has pointed upwind rather than downwind, and to this day winds are identified by the direction from which they blow rather than the direction in which they are blowing. For example, a westerly wind is a wind that blows from west to east, not one blowing in a westerly direction.

Each of the eight sides faces one of the principal wind directions, and there is a carved figure at the top of each face. The side facing north bears the figure of Boreas, the north wind, who is portrayed as a man wearing a cloak and blowing through a twisted seashell. Kaikas, the northeast wind, is a man carrying a shield from which he pours small, round objects that might be hailstones. Apeliotes, the east wind, is a young man holding a cloak filled with grains and fruit. Euros, the southeast wind, is an old man wrapped in a cloak. Notos, the south wind, is a man emptying an urn to produce a shower of rain. Lips, the southwest wind, is a boy pushing a ship. Zephyros, the west wind, is a young man carrying flowers. Skiron, the northwest wind, is a bearded man carrying a pot filled with charcoal and hot ashes. Triton pointed to the appropriate figure to tell people in the agora what weather to expect. The associations between particular wind directions and weather conditions are very local, of course. They apply only to Athens and at a time during a warm period that lasted from about 250–0 B.C.E.

Traditionally, the Greeks believed the weather was produced by the actions of the gods, and each god favored a particular type of weather. Athenians would have recognized the figures and known their meaning. The Tower was not referring to the gods, however,

The Tower of the Winds, in the Roman agora, Athens, Greece, is now a ruin, but when it was built, in about 100 B.C.E., it told Athenians what weather to expect. *(Dmitri Kessel/Stringer. Time Life Pictures)*

but to the fact that particular kinds of weather reach Athens from certain directions. It was based on observed facts, in the Aristotelian fashion.

As well as being a weather service, the Tower was also a public clock. Each side originally held a sundial, so throughout the hours of daylight at least one sundial would show the correct time—unless, that is, the sky was overcast. Not every day is sunny, but on cloudy days people could go inside the Tower to read the time from a clepsydra—a clock driven by falling water, which showed the time on a dial. In addition, the Tower had a disk that rotated, showing the movements of the constellations and the location of the Sun in relation to them. All in all, the Tower of the Winds was a remarkable scientific instrument.

JOSEPH HENRY, SAMUEL MORSE, AND THE TELEGRAPH

On May 1, 1844, the Whig Party held its national convention in Baltimore, Maryland, nominating Henry Clay (1777–1852) as its presidential candidate. News of the nomination was taken by hand to Alfred Vail (1807–59), who was waiting for it at Annapolis Junction. Vail transmitted it by electric telegraph to his colleague in Washington, Samuel Morse (1791–1872). This was the first news ever to be transmitted by wire. The telegraph wire was being installed between Washington and Baltimore, and Annapolis Junction was as far as it had reached on May 1. By May 24 it was completed, and on that day the message "What hath God wrought?" traveled in a fraction of a second from the U.S. Capitol in Washington, D.C., to Baltimore, Maryland, a distance of 40 miles (64 km).

The successful demonstration of the telegraph meant that for the first time in history information could travel over long distances faster than it could be carried by a horse-rider. Within a very short time telegraph lines were being erected across the United States and many European countries. By about 1854 there were 23,000 miles (37,000 km) of telegraph wires crisscrossing the United States. A submarine telegraph cable between England and France was laid in 1850, just six years after the first message was transmitted. An attempt was made in 1857 to lay a transatlantic cable, but the cable broke and the project was abandoned. A second attempt succeeded in 1866, allowing Europeans and Americans to exchange information in minutes, rather than the several weeks in each direction that the crossing took by ship.

Commerce, diplomacy, and journalism were the obvious beneficiaries, but they were not the only ones. Meteorologists could use the telegraph to assemble data gathered at widely dispersed observation stations, and they were quick to exploit the opportunity. Joseph Henry (1797–1878), the eminent American physicist, had been elected the first secretary of the Smithsonian Institution in 1846, and he used his position to set up a network of weather stations. By 1849, at a cost of $1,000, the Smithsonian Meteorological Project had recruited

150 volunteer observers. Ten years later there were 600 observers, some of whom were in Canada, Mexico, Central America, and the Caribbean, and by 1860 the annual budget for the project came to $4,000—a great deal of money in those days and equivalent to about $112,000 in today's money. At first most of the observers sent in their reports by mail, on standard forms supplied by the Smithsonian, but Henry had also persuaded several of the new telegraph companies to allow meteorological data to be carried free of charge, and by 1857 observers were able to take their reports to telegraph stations from New York to New Orleans.

Henry used the data he received to compile a daily weather map. The map was displayed in the Smithsonian building where the public had access to it. Each morning an assistant placed colored disks on the map. White disks indicated fine weather, black indicated rain, blue was for snow, and brown disks indicated cloudy conditions. Arrows on the disks showed the wind direction. Henry also shared the information with the *Washington Evening Star,* which began publishing daily weather reports for several cities in May 1857.

The Smithsonian weather map made it possible to forecast the weather, at least to a limited extent. Weather systems in middle latitudes travel from west to east, so imagining the colored disks shifted eastward gave a broad impression of the conditions people could expect. Henry was required to produce an annual report for the Smithsonian, and in his report for 1865 he called on the federal government to establish a national weather service that would issue forecasts and, in particular, warnings of approaching storms. Other influential people supported the idea, and in 1870 Congress passed legislation giving responsibility for weather forecasting to the Army Signal Service. In 1874 Henry persuaded the Signal Service to take over his volunteer observer scheme as well.

Inventions begin as ideas, but inventions of major historical importance such as the telegraph do not burst upon the world suddenly, from nowhere. The design that succeeds is the end product of a long process involving many scientists and engineers, and many failures. As early as the middle of the 18th century scientists knew that an electric current will travel a long distance along a wire provided one end of the wire is connected to the earth to complete the circuit. In 1753 an article by someone identified only as "C. M." appeared in the *Scots Magazine* suggesting the use of electric current to transmit information. C. M. proposed a bundle of insulated wires, one for each letter of the alphabet plus others for a space and essential punctuation. At the receiving end, each wire was connected to a ball and below each ball there was a piece of paper with the appropriate letter or symbol written on it. When current reached the ball the paper would jump up to it, and the message would be revealed letter by letter. Such a device would be very slow and cumbersome, however, and other inventors tried different approaches. Then, in 1825, the English inventor William Sturgeon (1783–1850) discovered that a needle enclosed in a coil of wire becomes a magnet when a current flows through the wire, and the magnet reverses polarity when the current reverses direction. He called this device an electromagnet. It offered the possibility of transmitting two signals, corresponding to reversals in current direction.

Joseph Henry made a signaling device based on the electromagnet. His apparatus rang a bell when a signal was received. Henry later invented the relay, which is a series of circuits connected so that when one is activated it activates the next in the sequence and the current flows through one circuit at a time. This overcame the diminution in a current flowing along a wire that is due to resistance in the wire. Relays made long-distance transmission feasible. Finally, Samuel Morse designed a telegraph that worked.

Morse also devised the code used to transmit messages. The telegraph could not convey voice messages or messages a printer could reproduce—the telephone came later and the teleprinter was introduced in the 1920s. The code Morse devised consisted of combinations of short and long impulses—dots and dashes. It was impressively simple, easy to learn, and remained in regular use by ships and aircraft until the 1940s, and it was 1995 before the U.S. Coast Guard abandoned it for communicating with ships at sea. The illustration shows an operator at Croydon Airport, near London, using the code to communicate with an aircraft. It is still used occasionally when radio conditions are poor, and many amateur enthusiasts use it. Morse's 1844 message was transmitted in Morse code to a receiver that embossed the sequence of dots and dashes on paper tape that moved along by a clockwork mechanism. Before long, however, operators learned to recognize the clicking sounds the receiving apparatus made and were able to transcribe these directly. The paper tape became obsolete.

A telegraph operator at Croydon Airport, near London, England, in 1928. He is using a key to transmit a message in Morse code to an aircraft. When the signaler on the aircraft replies the ground operator hears the code through his earphones and writes the message on paper. This drawing appeared in the *Illustrated London News* in 1928. *(Sheila Terry/Science Photo Library)*

Samuel Finley Breese Morse was born on April 27, 1791, at Charlestown, Massachusetts (now part of Boston). His father, Jedidiah (1761–1826) was a clergyman, educator, and also a geographer, whose portrait shows him behind a globe on which he rests his hand. Samuel was educated at Andover and Yale College (as it then was), where he studied religious philosophy and mathematics. While at Yale he developed a keen interest in painting, especially miniature portraits, and sold some of his work. After graduating in 1810 he persuaded his parents to allow him to travel to England to study historical painting.

He lived in England from 1811 until 1815 and gained admittance to the Royal Academy.

After returning to the United States, Morse settled in New York and prospered for a time as a portrait painter, although his historical paintings were never very popular. He was one of the founders of the National Academy of Design and from 1826 until 1845 he was its first president. He taught art at the University of the City of New York (now New York University), and he ran unsuccessfully for mayor of New York (on an anti-Catholic, anti-immigration ticket).

Morse spent the years from 1830 until 1832 traveling in Europe to improve his painting skills. It was during his return voyage that he met the American scientist and physician Charles Thomas Jackson (1805–80), who was carrying an electromagnet with him. The conversations he had with Jackson persuaded Morse that it might be possible to use electromagnets as the basis for a telegraph. He pursued this back in New York but soon found his knowledge of electricity was insufficient. A colleague advised him to talk to Joseph Henry (1797–1878). Henry had already devoted considerable thought to the problem and gave his advice freely, answering all of Morse's questions. By about 1835 Morse had developed a telegraph that would work. It took until 1844 for him to persuade Congress to fund its installation.

Once he had demonstrated the feasibility of the telegraph, Morse found himself caught up in a series of legal actions by others, including Jackson, who claimed priority for the invention. The illustration shows him at about this time. In 1853 he saw off the rival claims with a ruling from the Supreme Court but received no recognition from the federal government, despite the fact that they were using his invention. In 1858 the governments of several European countries made him a joint payment of 400,000 French francs (then worth about $80,000 and more than $2 million in today's money), but it was 1871 before Morse received any American recognition, when a statue of him was erected in Central Park, New York City. Nevertheless, by the end of his life

Samuel Morse (1791–1872), the inventor of the telegraph and of the Morse code, in a photograph taken on January 1, 1850. *(Hulton Archive/Getty Images)*

Morse was a wealthy man. He was generous and gave large sums of money to charity.

Samuel Morse married twice. His first wife, Lucretia Pickering Walker, died in 1825, following the birth of their third child. His second wife was Sarah Elizabeth Griswold. Morse died at his home in New York City on April 2, 1870. When it opened in 1900 on the campus of New York University, Morse was made a charter member of the Hall of Fame for Great Americans.

Joseph Henry was born in Albany, New York, on December 17, 1797. Both his parents, Ann Alexander Henry and William Henry, were immigrants from Scotland, who had arrived in the United States in 1775. Joseph's father worked as a laborer and died while Joseph was still young. The family was poor, and Joseph went to live with his grandmother in Galway, New York. He left school and at 13 was apprenticed to a watchmaker and silversmith in Albany, although at that time his first love was the theater and under different circumstances he might have become an actor. He might also have become a watchmaker, but for a curious event when he was 16 and spending a vacation on a farm owned by a relative. He was chasing a rabbit that disappeared beneath a church, and he crawled beneath the building to pursue it. Some of the floorboards were missing, and out of curiosity Henry climbed through the hole into the church. There he saw a shelf of books, one of which caught his eye. It was called *Popular Lectures on Experimental Philosophy, Astronomy, and Chemistry.* Henry read it and was inspired to enroll at Albany Academy, where students received the equivalent of a college education. He received free tuition, studying chemistry, anatomy, and physiology, and paid his way by tutoring and teaching in country schools, meanwhile reading widely on every aspect of science. In 1825 he obtained a job surveying the route for a new road to be built between the Hudson River and Lake Erie, in New York State. This aroused

The physicist Joseph Henry (1797–1878), one of the greatest of all American scientists, as he appeared in about 1860. As secretary to the Smithsonian Institution Joseph Henry established the network of weather observers that eventually grew into the National Weather Service. *(Hulton Archive/Stringer)*

an interest in engineering, but in 1826 he returned to Albany Academy as professor of mathematics and natural philosophy.

It was a busy life, but Henry managed to find time to experiment with electromagnets. He grew highly skilled at making them and he used them to make what he called "philosophical toys"—devices that foreshadowed the electric motor and telegraph. As his work became better known, Henry began to acquire an international reputation. In 1832 he was appointed professor of natural philosophy at the College of New Jersey (now Princeton University), a post that allowed him adequate time for research. He became secretary of the Smithsonian Institution in 1846, and from 1867 until he died he was the second president of the National Academy of Sciences. He helped organize the American Association for the Advancement of Science (publisher of the journal *Science*) and the Philosophical Society of Washington. The portrait shows him as he appeared in about 1860.

Joseph Henry died in Washington on May 13, 1878, and was buried in Oak Hill Cemetery. In his lifetime Henry was regarded as the greatest American scientist since Benjamin Franklin. Rutherford B. Hayes, president of the United States, attended his funeral, as did many other government officials.

CLEVELAND ABBE, FATHER OF THE WEATHER BUREAU

The *Washington Evening Star* began publishing weather reports supplied by the Smithsonian Institution in 1857, and other newspapers and news agencies soon followed. These reports gave details of conditions based on data from as few as 10 to about 50 locations. They were very local, and they did not include weather forecasts. That situation began to change on September 1, 1869, with the publication of the first *Weather Bulletin of the Cincinnati Observatory.* This was simply a report of weather conditions across the United States, but on September 22 the *Bulletin* included a three-day weather forecast, based on probabilities calculated by the forecaster. That forecaster was Cleveland Abbe (1838–1916), who had been appointed director of the observatory in 1868.

The first regular daily weather report with a three-day forecast appeared on February 19, 1871, and on November 8, 1871, the Cincinnati Observatory issued its first cautionary storm signal, warning of a risk of storms in the Great Lakes region. For the first six months of the new service Abbe prepared all the forecasts himself while he trained a team of assistants, and once his system was fully operating he insisted that the forecasts be verified by comparing them with the weather that actually occurred. During the first year they verified 69 percent of the forecasts, and Abbe apologized in the observatory's annual report for not checking all of them for lack of time. The forecasts were always based on probabilities, earning Abbe the nickname of Old Probability, often abbreviated to Old Prob.

Cincinnati Observatory was an astronomical observatory, but Abbe had widened its scope to include meteorology, which, after all, is very important to astronomers using telescopes. Weather forecasts were also of more general interest, and Abbe had managed to persuade the Cincinnati Chamber of Commerce, which funded the observatory, that Cincinnati was uniquely placed for issuing forecasts due to its location at the center of the national railroad and telegraph networks. In particular, Abbe emphasized the value of issuing storm warnings. What Abbe proposed, and what he wanted the Chamber of Commerce to finance, was to gather data by telegraph—Western Union Telegraph agreed to transmit weather reports free of charge—and study them at the observatory, then condense them into short, but informed and so far as possible accurate, two-day forecasts. The data would come daily from at least 100 stations. The forecasts would be passed to Associated Press for dissemination by telegraph to newspapers throughout the United States, and they would replace the many local reports. His scheme might appear visionary, but Abbe pointed out that it would not be the first, and he added force to his argument by warning that America might be falling behind. On July 29, 1868, in a letter to John A. Gano, the president of the Chamber of Commerce, he wrote:

> It cannot have escaped your notice that during the past 20 years very many endeavors have been made by various nations to utilize the science of meteorology. From the Paris Observatory daily bulletins are published showing the state of the weather in western Europe. In England storm warnings are published many hours in advance and sent to the ports that are threatened.

The growing interest in establishing a national weather service, and its obvious utility, soon attracted political support. On February 2, 1870, Representative Halbert E. Paine introduced a joint congressional resolution calling on the secretary of war to establish such a service. President Ulysses S. Grant signed the resolution on February 9, and the weather bureau came into being in November 1870, forming part of the recently formed division of telegrams and reports for the benefit of commerce of the Army Signal Service. The bureau was headed by General Albert J. Myer, the chief signal officer, and in January 1871 Cleveland Abbe was appointed as a civilian assistant. It was Abbe who prepared the weather reports and forecasts. In 1873 Abbe launched the *Monthly Weather Review* and his advice helped persuade General Myer to seek the collaboration of the International Meteorological Congress of 1873 and of foreign governments in publishing a *Daily Bulletin of Simultaneous International Meteorological Observations.*

With observers reporting from weather stations across the country Abbe recognized a need to standardize the times they recorded their measurements, because each community was using its own local time. He divided the country into four standard time zones, and in 1883 he persuaded the railroad companies to adopt his system, which greatly simplified the preparation of timetables. Abbe's system is the one that was eventually adopted in August 1916.

In 1891 the weather bureau was transferred from the army to the Department of Agriculture and renamed the United States Weather Bureau (it was transferred to the Department of Commerce in 1940). Cleveland Abbe was the meteorologist in charge, a position he retained until almost the end of his life. At the same time, he continued his own research and taught meteorology at Johns Hopkins University.

Cleveland Abbe was born in New York City on December 3, 1838. His father, George Waldo Abbe, was a merchant. His mother was Charlotte Colgate Abbe. Abbe attended several schools in New York, and in 1851 he enrolled at the New York Free Academy (now City College of the City University of New York), graduating in 1857 and obtaining a master's degree in 1860. He taught mathematics at Trinity Latin School, New York, in 1857 and 1858, and in 1859 obtained a post as assistant professor of engineering at Michigan Agricultural College. Later the same year he became a tutor in engineering at the University of Michigan, where he also studied astronomy.

In April 1860 President Lincoln called for volunteers to join the Union army. Abbe responded, but was rejected because of his severe myopia. Instead, he spent the years from 1860 until 1864 assisting the astronomer Benjamin Apthorp Gould (1824–96) at the United States Coast Survey, in Cambridge, Massachusetts. Abbe spent 1865 and 1866 as a guest astronomer at the Nicholas Central Observatory at Pulkova near St. Petersburg, Russia, an institution he later described as a scholar's paradise. He returned to the United States in 1867 and worked for a short time at the U.S. Naval Observatory before becoming director of the Cincinnati Observatory.

Abbe married twice, first in 1870 to Frances Martha Neal. They had three sons. Frances died in 1908 and in 1909 Abbe married Margaret Augusta Percival. Cleveland Abbe worked for the Weather Bureau until 1915, when failing health made it impossible for him to continue. He retired to Chevy Chase, Maryland, where he died on October 28, 1916.

ROBERT FITZROY AND THE FIRST NEWSPAPER WEATHER FORECAST

Cleveland Abbe had drawn the attention of the Cincinnati Chamber of Commerce to the fact that America was in danger of falling behind in the relatively new science of meteorology. What he did not mention was that on August 31, 1848, the first daily newspaper weather report had appeared in the *Daily News,* published in London. The paper had commissioned James Glaisher (see pages 136–138) to collect meteorological data from around Britain. In 1860 the *Times* of London published the first daily weather forecast, which had been issued by the office of the Meteorological Statist to the Board of Trade. A statist would now be called a statistician, and the meteorological statist, with a staff of three, was Vice Admiral Robert FitzRoy, FRS (Fellow of the Royal Society; 1805–65). The photograph of FitzRoy (page 189) was taken in about 1850, the year he retired from active service.

Robert FitzRoy was captain of HMS *Beagle* during the five-year voyage around the world on which Charles Darwin (1809–82) sailed as a naturalist and as a companion who could dine with the captain and who shared his interest in science. FitzRoy's main interest was in meteorology, and in the course of the voyage he designed a *storm glass.* This device was popular in the 18th and 19th centuries. It consisted of a tightly sealed glass cylinder containing a mixture of chemical compounds. Crystals formed and dissolved inside the cylinder, apparently in response to meteorological changes, and the storm glass was used to predict the weather. The government later distributed these devices, known as FitzRoy's storm barometers, to fishing ports around Britain. Owners of fishing fleets were forbidden to send boats to sea when the barometers warned of a storm. They made FitzRoy a hero among the fishermen, but he was highly unpopular with the owners.

His interest in weather and his deep concern for the safety of seafarers led to FitzRoy's appointment in 1854 to head a new department that would collect meteorological data from ships at sea. He obtained suitable instruments that had been tested for reliability and these were loaned to sea captains. By May 1855, 50 merchant ships and 30 naval vessels were equipped with them. FitzRoy also developed a mercury barometer that was produced in sufficient numbers for them to be installed in stone huts at fishing ports around Britain. (A few of the stone huts still exist.) Some of the barometers (these were distinct from his storm glasses) had thermometers and storm glasses mounted beside them. In 1858 FitzRoy

Robert FitzRoy (1805–65) was a British naval officer best known as the captain of HMS *Beagle*. At the end of his active naval service, however, FitzRoy embarked on a second career as the head of what later became the UK Meteorological Office. *(Hulton Archive/Getty Images)*

published a 50-page book of instructions for interpreting readings from the barometers.

In October 1859 the iron, steam-powered clipper *Royal Charter* sank in a storm off the coast of Anglesey, in North Wales, with the loss of 450 lives, but she was not the only victim. The Royal Charter Storm, as it came to be called, damaged or sank more than 200 ships. It was one of a series of storms that continued through October and November, claiming a total of 325 ships and 748 lives. Robert FitzRoy was convinced that the disaster could have been averted or its severity reduced if only adequate warnings had been issued. In June 1860 FitzRoy was given permission to install a warning system, and by September his barometers and other instruments had been distributed to 18 ports, and local telegraph operators had been

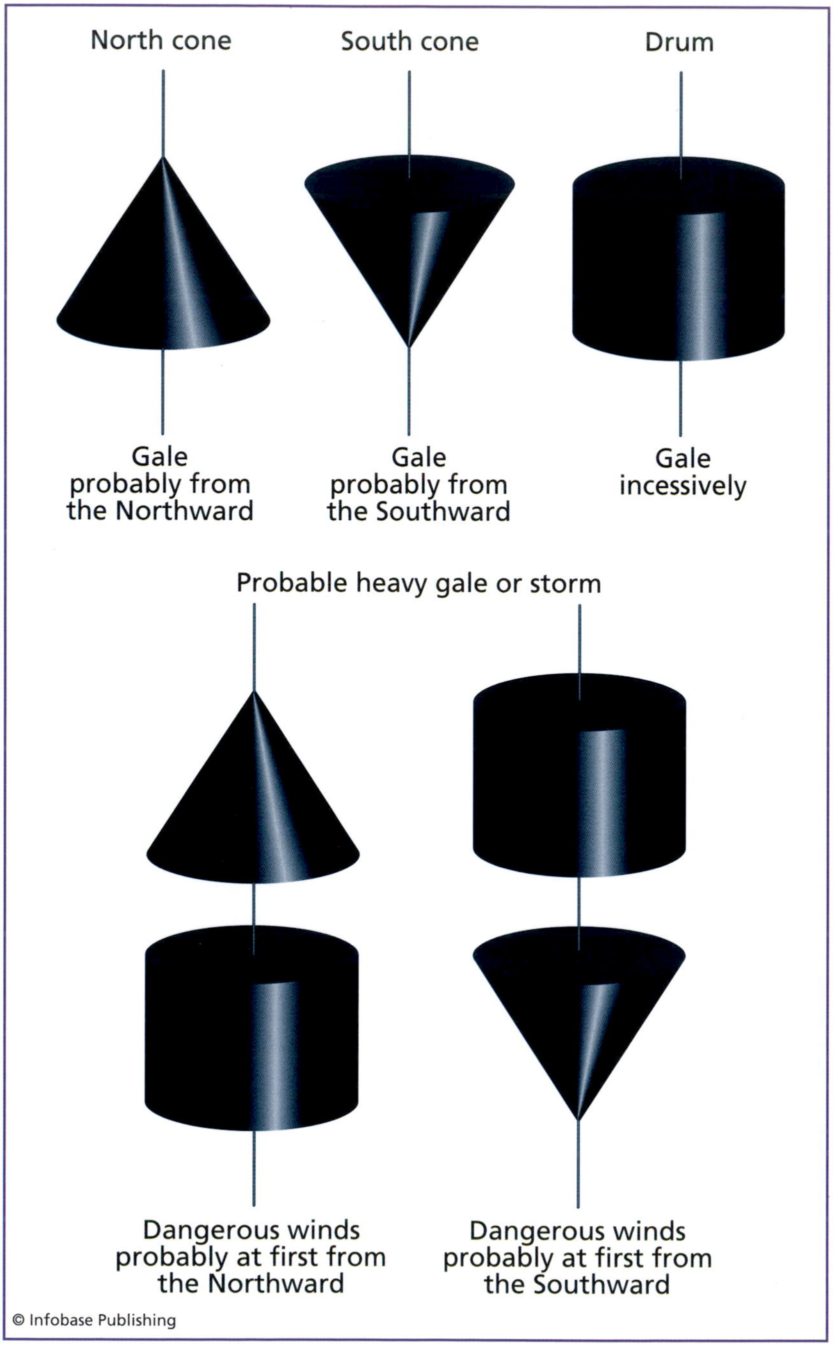

North cone   South cone   Drum

Gale
probably from
the Northward

Gale
probably from
the Southward

Gale
incessively

Probable heavy gale or storm

Dangerous winds
probably at first from
the Northward

Dangerous winds
probably at first from
the Southward

© Infobase Publishing

The system of cones and drums (cylinders) devised by Robert FitzRoy to warn of approaching storms consisted of cones and drums hoisted on masts 30–40 feet (9–12 m) high situated on the coast, where they were clearly visible from all directions. At night signal lamps were used.

commissioned to send the instrument readings to FitzRoy's London office. He also introduced a set of warning signals. These consisted of cones and drums (cylinders) displayed from tall masts by day and signal lights by night. The following illustration shows how they were used, based on the drawing in *The Weather Book: A Manual of Practical Meteorology* by Rear Admiral Robert FitzRoy, published in 1863. They were so famous that the expression "raising a storm cone" became a popular metaphor for warning of impending disaster.

Robert FitzRoy was born on July 5, 1805, at Ampton Hall, in Suffolk, England. His father was General Lord Charles FitzRoy, a descendant of King Charles II and at one time an aide-de-camp to King George III. His mother was Lady Frances Anne Stewart, the daughter of the Marquis of Londonderry. The family belonged to the highest ranks of the English and Irish aristocracies. When he was 12 Robert entered the Royal Naval College at Portsmouth. He graduated in 1819 and entered the Royal Navy. Having achieved marks of 100 percent in the qualification examination, FitzRoy was commissioned as a lieutenant on September 7, 1824. He served on HMS *Thetis* and in 1828 on HMS *Ganges* as flag lieutenant to Rear Admiral Sir Robert Waller Otway. Later that year the captain of HMS *Beagle* died, and its first officer brought the ship to Rio de Janeiro. The *Beagle* had been engaged on a hydrographic survey of Tierra del Fuego, and on December 15 Admiral Otway appointed FitzRoy as its captain. FitzRoy continued with the mission, returning on October 14, 1830.

In 1831 FitzRoy stood for election to parliament, but was defeated. He seemed to have no prospect of any further command, but his friend Francis Beaufort, head of the hydrographic office at the Admiralty (see "Francis Beaufort and His Wind Scale" on pages 118–122), and his uncle the duke of Grafton spoke on his behalf, and FitzRoy was reappointed to the command of the *Beagle.* Aware of the length of the voyage he was about to undertake and the fact that his predecessor as captain of the *Beagle* had committed suicide in a fit of depression, he asked Beaufort to recommend someone who might accompany him as a companion. Beaufort made enquiries, and Charles Darwin was appointed.

The *Beagle* returned to England on October 2, 1836. Later that year FitzRoy married May Henrietta O'Brien; they had three daughters and one son. Following Henrietta's death, in 1854 FitzRoy married Maria Isabella Smith; they had one daughter. FitzRoy devoted himself to writing an account of the voyage. *Narrative of the Surveying Voyages of His Majesty's Ships* Adventure *and* Beagle *Between the Years 1826 and 1836, Describing Their Examination of the Southern Shores of South America, and the* Beagle's *Circumnavigation of the Globe,* by Robert Fitzroy, was published in 1839 as two volumes. Darwin's *Journal and Remarks* (later known as *Voyage of the* Beagle) formed a third volume, and a book-length appendix to volume 2 formed a fourth volume. The Royal Geographical Society awarded FitzRoy a gold medal in 1837.

In 1841 FitzRoy was elected member of parliament for Durham. The following year the first governor of New Zealand died, and on the recommendation of the Church Missionary Society FitzRoy was appointed to succeed him. He took up the post in December 1843 with instructions to maintain the peace, protect the Maori people, but allow the settlers as much land as they required. He was allowed very little money to achieve this and very few troops. It was an impossible task, and the New Zealand Company, representing the interests of the settlers, was influential in having him dismissed in 1845, mainly because he considered the Maori claims to the land as valid as those of the settlers.

FitzRoy returned to England in 1848 and, far from being disgraced, was made superintendent of the Royal Naval Dockyards at Woolwich. In 1848 he was given his last command, of HMS *Arrogant.* He retired from active service in 1850, due partly to ill health, and for a short time in 1853 he was secretary to his wife's uncle, Lord Hardinge, commander in chief of the army. In 1851 FitzRoy was made a fellow of the Royal Society with the support of 13 fellows, including Charles Darwin. In 1854, on the recommendation of the president of the Royal Society, FitzRoy took up his post as meteorological statist to the Board of Trade. He retired in 1863, with the rank of Vice Admiral.

Robert FitzRoy was prone to depression, and criticism hurt him badly. Newspapers and some politicians disliked his humanitarian views, and the American naval officer and meteorologist Matthew Maury (1806–73) disapproved of his scientific methods. His deepest pain, however, came from the conflict between his religious views and the challenge Darwin's work presented to them, compounded by the guilt he felt at having helped Darwin formulate his theory. On April 30, 1865, Robert FitzRoy took his own life.

FRANCIS GALTON AND THE FIRST WEATHER MAP

On November 14, 1854, during the Crimean War of 1853–56, a fierce storm devastated the Anglo-French fleet in the Black Sea near the town of Balaklava, and the French minister for war asked Urbain-Jean-Joseph Leverrier (1811–77) to study the storm to see if lessons could be learned from it. Leverrier, a distinguished astronomer, had established weather stations throughout France and had issued the first weather reports for a number of French cities. He used reports from weather observers to determine the track of the Balaklava storm and was able to show that if these data had been plotted on maps and the necessary information telegraphed to naval commanders in threatened ports the disaster might have been averted.

In England, Francis Galton (1822–1911) knew of Leverrier's work, and also that of Robert FitzRoy (1805–65; see "Robert FitzRoy and the First Newspaper Weather Forecast" on pages 188–192) and Matthew Maury (1806–73), the American meteorologist and oceanographer who was exploring the links between weather systems and ocean currents. Galton sent a detailed questionnaire to weather stations throughout the British Isles, asking for information about the weather during the month of December 1861. When the replies reached him he plotted the data on a map, using symbols he had invented for the purpose, and by 1862 he had succeeded in drawing the world's first detailed weather map. His map revealed a feature that no one previously had suspected. Other meteorologists had established that air moves in a counterclockwise direction around centers of low atmospheric pressure (see "Christoph Buys Ballot and His Law" on pages 122–125), but what Galton discovered was the opposite: Air circulates in a clockwise direction around centers of high pressure. He coined the term *anticyclone* to describe an area of high pressure.

Galton's next step was to find a way for newspapers to print weather maps. He modified the pantograph, an instrument used for copying drawings, so it would incise lines in a soft material from which printing blocks could be made. He also designed typefaces, including his weather symbols, which could be printed onto the maps. *The Times* used Galton's techniques in preparing the weather maps it printed from data supplied by the Meteorological Office.

Galton also studied winds in the upper atmosphere, but rather than use kites or balloons, he used shells that released smoke when they exploded. Working in close collaboration with scientists from the Meteorological Office, Galton had shells made that were designed to explode at precisely 9,000 feet (2,745 m). The tests were carried out over a section of the Irish Sea where falling debris was unlikely to damage ships, and they were very successful. It was easy to track the movement of the smoke. Responding to a suggestion by Robert Fitz-Roy, Galton also devised the diagram, called a *wind rose*, that shows the direction of the prevailing winds. The illustration shows what a wind rose looks like and explains how it is prepared.

Francis Galton was born on February 16, 1822, in Sparkbrook, then a rural village and now a district of Birmingham, England, into a Quaker family of prosperous gun manufacturers and bankers. He was the youngest of nine children and a cousin to Charles Darwin

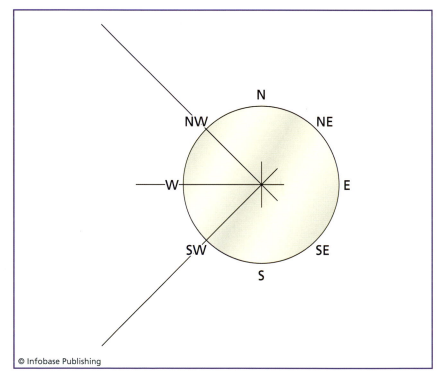

Wind direction is recorded throughout a month, and the daily wind directions are grouped into the eight compass directions. The wind direction on each day is represented by a line of a convenient unit length drawn in that direction from the center of a circle. If the wind blows from the same direction on another day the line is extended by one unit length. In this example the prevailing winds are from the southwest and northwest.

(1809–82). Francis was a child prodigy. He could read before the age of three, was studying Latin at four, knew some Greek and could perform long division at five, and by the time he was six he was reading adult books for pleasure, including Shakespeare. He attended several schools before entering medical school at his parents' suggestion. He studied for two years at Birmingham General Hospital and King's College Hospital in London. Then, from 1840 until 1844, he studied mathematics at Trinity College, University of Cambridge. A severe mental breakdown prevented him from taking an honors degree, and he graduated with a pass degree. He returned briefly to his medical studies, but ended them when his father died in 1844, leaving him a substantial inheritance. With no need to earn a living, Galton could pursue whatever topic appealed to him.

His first love was travel, and in 1845 and 1846 Galton visited Egypt and followed the Nile upstream to Khartoum, then traveled to Beirut, Damascus, and down the Jordan River. He joined the Royal Geographical Society in 1850, and spent 1850 and 1851 on a journey of 1,700 miles (2,735 km) through southwestern Africa (now Namibia). In 1853 the Royal Geographical Society awarded him its gold medal and the French Geographical Society its silver medal for his work mapping this little-known region. Also in 1853 Galton married Louisa Jane Butler. They had no children.

Galton was about to become a successful author. His advice to Victorian travelers, *The Art of Travel; or, Shifts and Contrivances Available in Wild Countries,* appeared in 1855, published in London by John Murray, and was a best seller. It went through five editions, the last in 1872, and that edition is still in print. His African travels had seriously damaged his health, however, and although Galton continued to travel in Europe, he never returned to Africa.

Francis Galton was a polymath. As well as his contributions to meteorology and exploration, Galton developed new statistical techniques. He was the first person to show that fingerprints are unique, and he partly devised a method for identifying them. He invented a teletype printer and the ultrasonic dog whistle and a word-association test that was widely used.

When Darwin's *On the Origin of Species by Means of Natural Selection* appeared in 1859, Galton's life changed, and from then he devoted most of his time and energy to the study of heredity and in particular to the question of whether human intellectual ability was inherited. He was the first person to use the expression *nature and nurture.* This work led to his suggestion that it might be possible to improve the human race by selective breeding, an idea known as eugenics—a word Galton coined—that is now discredited.

Galton was elected a fellow of the Royal Society in 1860 and in 1910 received its Copley Medal. He was knighted in 1909. Sir Francis Galton died at Haslemere, Surrey, on January 17, 1911.

LEWIS FRY RICHARDSON, FORECASTING BY NUMBERS

Vilhelm Bjerknes (1862–1951) and his colleagues in Bergen (see "Vilhelm Bjerknes and the Bergen School" on pages 149–154) had suggested the possibility of forecasting the weather mathematically. The alternative is to estimate the speed and direction known weather systems are moving and calculate when the weather associated with them will reach the places the forecast is meant to serve. To a mathematician, numerical forecasting as it is called must seem more attractive and more reliable. Certainly that was the view of Lewis Fry Richardson (1881–1953), an English mathematician and meteorologist who devised ways to prepare forecasts in this way. He described his method in 1922 in a book called *Weather Prediction by Numerical Process.* The book aroused little interest and was soon forgotten.

The principle appears straightforward. A great deal is known about the way the air behaves under different conditions (see chapter 4, "How Gases Behave" on pages 100–113). Richardson believed that meteorologists understood the physical laws governing the atmosphere sufficiently well to be able to use them as a basis on which to predict the future state of the atmosphere provided an earlier state was known. His idea was to mark the location of every weather station in the world on a large map. Data from each of these stations describing the state of the atmosphere—temperature, pressure, and humidity from the surface to the tropopause, cloud type and extent, precipitation, and the rate of change in these factors—would be passed to teams of operators sitting in a large hall. For convenience, the teams of operators, or computers as Richardson called them, would be seated in a pattern matching the distribution of their weather stations, so the hall would resemble the map. Each operator would apply to the data a differential equation or part of one equation worked out by Richardson. This would generate new data affecting neighboring stations. A supervisor seated in a large pulpit overlooking the hall would coordinate the work of the operators, shining a red light on those whose work was running ahead of the others and a blue light on those who were falling behind. And from time to time fresh data would arrive from the real weather stations. The supervisor had assistants and messengers, and there were four senior clerks who collected the details of the future weather as fast as it was being computed and sent it by pneumatic tube to a quiet room where it would be coded and telephoned to a radio transmitting station to be broadcast.

It was a huge concept. In his 1922 book Richardson described the hall:

> Imagine a large hall like a theater, except that the circles and galleries go right round through the space usually occupied by the stage. The walls of this chamber are painted to form a map of the globe. The ceiling represents the north polar regions, England is in the gallery, the tropics in the upper circle, Australia on the dress circle and the Antarctic in the pit. . . . From the floor of the pit a tall pillar rises to half the height of the hall. It carries a large pulpit on its top. In this sits the man in charge of the whole theatre . . .

Despite the scale of the operation, Richardson tested it by himself, using data obtained from records to forecast mathematically the weather on a particular day—he chose May 20, 1910—and compare his forecast to what actually happened. He was spectacularly wrong, predicting that the air pressure would increase by two pounds per square inch (14.5 kPa) in six hours when in fact the pressure barely changed. More detailed analysis of his calculations, however, showed that had he applied techniques to smooth out the data his result would have been fairly accurate. It was an outstanding achievement, given that Richardson performed all the calculations with only a slide rule to help him—in his day computer and calculator both meant people who performed computations and calculations—and he did the work while serving in an ambulance unit in France during World War I.

Richardson estimated that approximately 64,000 operators equipped with slide rules would work in his hall, or theater. It is more likely, however, that to monitor weather all over the world, as he proposed, would have required more than 200,000 operators. Even then, they would have been able to do no more than keep up with the weather as it happened. If they were to work faster than the real weather to produce a forecast the hall would have had to hold approximately 1 million operators and coordinating their work might have called for more than a single supervisor and his assistants overseeing them from a pulpit.

It is hardly surprising that meteorologists remained unimpressed by this method of forecasting, and the first edition of his book sold only 750 copies. Views began to change with the introduction of electronic computers, however. Richardson's book was republished in 1965, sold thousands of copies, and is still in print. Meteorologists now have access to very large supercomputers that are able to handle the volumes of data numerical forecasting requires and that can perform calculations at a speed high enough to produce useful forecasts. They also have access to data from the upper atmosphere, which was not available to Richardson, and advances in knowledge have allowed scientists to improve some of his equations.

Lewis Fry Richardson was born on October 11, 1881, in Newcastle-upon-Tyne, an industrial city in the northeast of England. He was the youngest of the seven children of David Richardson (1835–1913), who ran a business tanning leather, and Catherine Fry (1838–1919). They were Quakers and a prosperous family. Lewis attended Newcastle Preparatory School and in 1894 he entered Bootham School in York, a school for Quakers, where he was introduced to meteorology and where his Quaker beliefs were reinforced. In 1898 he moved to Durham College of Science (now Newcastle University) to study mathematics, physics, chemistry, botany, and zoology. He graduated with a first-class degree in 1903 and obtained a job at the National Physical Laboratory, where he worked in 1903 and 1904, and again from 1907 to 1909. He held posts at University College Aberystwyth in 1905–06 and Manchester College of Technology in 1912–13. He also worked at National Peat Industries (1906–07) and the Sunbeam Lamp Company (1909–12). In 1913 he went to work at the Meteorological Office as superintendent of Eskdalemuir Observatory in southwestern Scotland.

When World War I began in 1914 Richardson declared himself a conscientious objector in accordance with his Quaker pacifism, but he did not remain outside the war. From 1916 until 1919 he served in the Friends' Ambulance Unit, funded by the Quakers and attached to the 16th French Infantry Division. His service in France allowed him to exercise another passion, for Esperanto, which he tried out on German prisoners of war. It was while he was in France that he worked out the equations needed for weather forecasting. His notes were lost, but later found beneath a pile of coal.

After the war Richardson returned to the Meteorological Office, but in 1920 the Meteorological Office was transferred to the Air Ministry. Richardson's pacifist beliefs would not permit him to work for the military, so he resigned. From 1920 until 1929 he was head of the physics department at Westminster Training College and from 1929 until 1940 he was principal of Paisley College of Technology and School of Art (now the University of Paisley) in the west of Scotland. He retired in 1940 and went to live in the village of Kilmun, Argyll.

Retirement gave Richardson time to develop another of his interests, which was the application of mathematics to the causes of war. This led to a psychological theory of war, expressed mathematically, and Richardson wrote a number of papers and several books on the topic, including *Generalized Foreign Politics* (1939), *Arms and Insecurity* (1940), and *Statistics of Deadly Quarrels* (1950). In the course of this work he noted the difference in measurements of national frontiers quoted by different authorities. He realized that the length of a frontier or coastline depended on the ruler used to measure it, and as the size of the ruler decreased the length increased without limit. This is now called the Richardson effect, and it helped generate the branch of mathematics known as fractals.

Richardson was best known during his lifetime for his studies of atmospheric turbulence. He was elected a fellow of the Royal Society in 1926. He died at Kilmun on September 30, 1953.

EDWARD LORENZ, CHAOS, AND THE BUTTERFLY EFFECT

Numerical forecasting is now a standard technique that meteorologists use to help them compile their forecasts. They also have satellite images and data, sent to their computers in real time, and historical records of the kind of weather that often follows the atmospheric situation they are currently observing. Modern weather forecasting is a highly technical, highly mathematical process. Even so, since the 1960s scientists have known that it is limited, because it is impossible to predict the weather very far ahead in middle latitudes, where conditions are very changeable. Detailed forecasts are reliable for a few days, general forecasts for rather longer, but if someone wishes to know whether the weather will be fine enough for an outdoor party on a date months ahead, no one can tell them. It may be that forecasting the weather months ahead will forever remain impossible. The reason is that weather systems behave chaotically. Chaos is a branch of mathematics, and it was a meteorologist who discovered it.

One day in 1961, Edward Lorenz (1917–2008), a meteorologist working at the Massachusetts Institute of Technology (MIT), was using a computer model to help him understand how weather happens. A computer model is a set of equations, similar to those used in numerical forecasting, that simulate the way the atmosphere responds to changes. The scientist constructing the model places an imaginary grid over the area of interest, which may be the entire Earth. The grid has several levels from the surface to the tropopause, giving it a three-dimensional structure. The equations are then applied at every point where grid lines intersect. There are many such points, and several calculations at each. It is a formidable computing task. Lorenz had made his own model—one of the first—using 12 differential equations. When he ran the model it described the way a weather system evolved over time. This was not a forecast of real weather, but the weather it described resembled real weather. The computer presented its results as a paper printout.

On that particular occasion Lorenz wished to repeat a particular sequence of weather. Rather than run the model from the beginning he started it in the middle, keying in the data the computer had printed out at the relevant place. He entered the data, set the model running, and left it for an hour. When he returned he found that the computer had not repeated the earlier sequence as he expected. The new sequence began in the same way, but it quickly began to diverge from the earlier sequence until the two were totally different. The computer had used the same data and the same equations to produce two entirely different weather systems.

Lorenz set to work to discover what had happened and eventually he succeeded. As it ran his model, the computer stored the results of its calculations in its internal memory to six decimal places, but to save paper Lorenz had programmed the computer to print out its data to only three decimal places, so, for example, the number 0.794205 in the memory printed out as 0.794, and 0.794 was the number Lorenz keyed in for the second run. The difference between them was minute and should have made no difference to the outcome. But the difference those missing decimal places made was huge.

In the years that followed Lorenz continued to work on his discovery and on December 29, 1979, he described it in a paper he read at the annual meeting of the American Association for the Advancement of Science, held that year in Washington, D.C. The title of his paper was "Predictability: Does the Flap of a Butterfly's Wings in Brazil Set Off a Tornado in Texas?" It came to be known as the *butterfly effect,* and the mathematics underlying it gave rise to chaos theory. Despite its name, a chaotic system is not disordered. Rather, it displays an order of a kind that was unknown until Lorenz's computer model revealed it. Previously, mathematicians had believed there were two kinds of order: steady state and periodic. The variables in a steady state system do not change over time. Variables in a periodic system do change, but in a regularly repeating fashion that forms an endlessly repeating loop when plotted on a graph. Changes in a system that changes chaotically can also be plotted on a graph, but those changes follow a double spiral path that never repeats itself, drawing closer and closer to a point known as a chaotic attractor but never reaching it. The behavior is ordered, but because it does not repeat it is impossible to predict its state at any particular point in the future. In the case of the weather, the seasons follow each other, but it is impossible to predict the date on which the first spring flowers will open next year or the date of the first fall of snow.

The implications are profound. A scientist's initial response might be to seek more data, to reduce to a minimum the small variations between the computer model and the real weather outside. But that is impossible. In the example above, the difference between 0.794205 and 0.794 is 0.000205. No measurements made in the real world, of temperature, pressure, humidity, or any other variables, can come close to that degree of accuracy. Random air movements, minute differences between two instruments, a difference of seconds in the time of day, and countless other tiny variations add up to what is known as noise. Noise can be reduced, but never eliminated, and it means that any measurement is, to some extent, a statement of probability.

Consequently, it is impossible to specify the initial conditions of the atmosphere with the precision needed to predict mathematically how a weather system will develop over more than a few days. It is also impossible for a computer model to predict for decades or centuries ahead how the climate of a region or the world will respond to small changes in the state or composition of the atmosphere.

Edward Norton Lorenz was born on May 23, 1917, in West Hartford, Connecticut. He studied mathematics at Dartmouth College, Harvard University, and MIT. During World War II he served as a weather forecaster for the U.S. Army Air Corps. In 1946 he obtained a faculty position at MIT, where he also obtained his doctorate of science in meteorology in 1948. From 1962 until 1987 he was professor of meteorology at MIT and was then professor emeritus of meteorology. He died on April 16, 2008, at his home in Cambridge, Massachusetts.

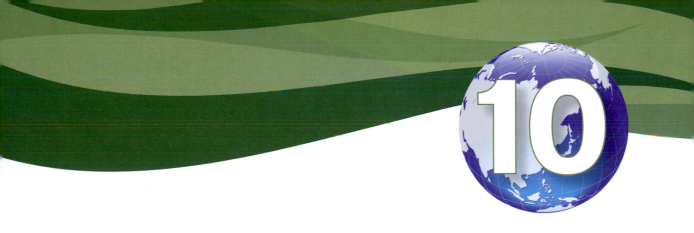

Changing Climates

As the seasons follow one another year after year, the everyday weather undergoes predictable changes. Some years are warmer than others, or cooler. There are good summers, with long, sunny days, and miserable summers when it seems to rain for most of the time. But these are memories, and human memory is notoriously unreliable. We remember the exceptions and forget the ordinary. Elderly people remember the hot summer days and winter snows of their childhood, but those are remarkable, and ordinary summers and winters, when the weather is not unusually hot or cold, provide nothing on which memory can seize, no handle to open memory's door. When we allow for the way memory works, it appears that the weather averages out over the years and that the climate remains constant. Climate is the average of the weather conditions experienced in a particular place over a long period, commonly 30 years.

It is true that climates change little over a human lifetime. It is far from true that they do not change over longer periods. This final chapter describes how scientists revealed two long-term changes and the processes that cause climatic change.

LOUIS AGASSIZ, JEAN CHARPENTIER, AND THE ICE AGE

In 1815 Jean-Pierre Perraudin (1767–1858), who hunted chamois in the Swiss Alps and knew the mountains very well, became puzzled by giant granite boulders he saw perched high in mountain valleys.

These rocks were of a quite different type than the solid rocks on which they rested. The boulders were obviously extremely heavy, and it would have required much effort to carry them to the places where he found them. He did not believe a great flood could have carried them. So how did they come to be there? Perraudin also noticed long, parallel grooves scratched on the rocks lining the valleys. He suspected *glaciers* were responsible, but had no idea how that might be.

Perraudin sought the opinion of one of the most famous geologists of the time, Jean de Charpentier (1786–1855). Charpentier simply dismissed as outrageous the idea that glaciers could have transported boulders and scratched grooves on rocks. One person was persuaded, however. Ignatz Venetz (1788–1859) was a Swiss civil engineer who had recently moved to the area. Perraudin convinced Venetz, and in 1829 a lecture on the subject delivered by Venetz succeeded in convincing Charpentier. Charpentier then approached Louis Agassiz (1807–73). Agassiz was professor of natural history at the University of Neuchâtel and famous for his studies of fossil fish, and it was important to win his support if the glacier theory were to gain wide acceptance. Agassiz was not impressed, but Charpentier persisted and in 1836 persuaded Agassiz to accompany him on a visit to the mountains and look at the evidence. Still skeptical, Agassiz agreed, but when he examined the rocks he was intrigued. He firmly believed in the biblical flood and its power to transport large rocks, but he recognized the strength of the glacial theory and resolved to test it.

In 1836 and 1837 Agassiz spent his summer vacations with friends on the Aar Glacier, staying in a hut they built and which they called the "Hôtel des Neuchâtelois." They studied the rocks on either side of the glacier and drove a straight line of stakes into the ice from one side of the glacier to the other. By 1841 the line was no longer straight: The central part formed a bulge, demonstrating that the ice had moved. In 1839 Agassiz came across another hut that had been erected in 1827 and that had moved 1 mile (1.6 km) from its original position. Then Agassiz widened his search to boulders that littered the plain of eastern France and that were quite different from the underlying rock. He found that these boulders matched rocks in Scandinavia.

A new theory was forming, even more radical than Charpentier's idea that glaciers had once extended beyond their 19th-century positions. Agassiz had come to believe that at one time ice had covered all the land from the North Pole to the Mediterranean. On July 24, 1837, he introduced this idea in his presidential address to the Société Helvétique de Sciences Naturelles (Swiss Society of Natural History), meeting in Neuchâtel. In 1840 Agassiz described his researches and conclusions in a book, *Études sur les glaciers* (Studies of glaciers).

Agassiz's theory of what he called the Great Ice Age that had gripped the Earth quite recently in geological terms gradually won acceptance, but it also aroused controversy. There was great debate— indeed, heated argument—about how glaciers move. A distinguished Scottish physicist James David Forbes (1809–68) proposed that under sufficient pressure ice becomes plastic and will flow. The Irish physicist John Tyndall (1820–93) disagreed, asserting that ice cannot flow plastically, but must repeatedly thaw and refreeze, a process called *regelation.* James Thomson (1822–92), the elder brother of Lord Kelvin (1824–1907), joined in, suggesting that the melting point of ice decreases as pressure on the ice increases, and that pressure-melting at the base lubricates the glacier, and that is how a glacier flows.

Whatever the mechanism of flow—and the view today is that *valley glaciers* move through a combination of regelation and plastic flow and that pressure-melting affects thick ice sheets—Agassiz continued to find evidence to support his theory of a single ice age affecting most or all of the Northern Hemisphere. He found evidence of glaciation in Scotland and in North America, which he visited in 1846. Scientists now know that there have been many ice ages and that Agassiz had discovered evidence of the most recent, which ended approximately 10,000 years ago.

Jean-Louis-Rodolphe Agassiz was born on May 28, 1807, at Motier, Switzerland, the son of a Protestant pastor. He attended school in Bienne, northwest of Bern, and Lausanne. In 1824 he enrolled at the University of Zürich and in 1826 transferred to the University of Heidelberg, Germany. He caught typhoid in Heidelberg, however, and was compelled to return home to recuperate. In 1827 he enrolled at the University of Munich. He obtained his Ph.D. in 1829 from the University of Erlangen and qualified as a doctor of medicine at the University of Munich in 1830.

In 1826 Agassiz was given the job of classifying specimens of fish collected on a two-year expedition to the Amazon. His classification was published in 1829, and its success determined the course of Agassiz's career. In 1831 he moved to Paris to continue his work with fishes at the Natural History Museum under the supervision of Georges Cuvier (1769–1832). Following Cuvier's death, Alexander von Humboldt (1769–1859) helped Agassiz obtain the professorship of natural history at the University of Neuchâtel.

In 1846 a grant from King Wilhelm Friedrich IV of Prussia paid for Agassiz to visit the United States, partly to continue his studies and partly to deliver a series of lectures at the Lowell Institute in Boston. The lectures proved popular and Agassiz extended his stay. In 1848 he was appointed professor of zoology at Harvard University, where he remained for the rest of his life, becoming an American citizen. In 1858 he helped establish the Museum of Comparative Zoology at Harvard, building it around his own collection, and he was director of the museum from 1859 until his death. The illustration shows Agassiz when he was about 43 years old. Louis Agassiz was one of the finest science teachers America has ever known. He was devoted to his students, treating them as colleagues.

Agassiz married twice. His first wife, Cécile Braun (1808–48), died at Baden. In 1850 he married Elizabeth Cabot Cary (1822–1907), who became his scientific assistant. Louis Agassiz died on December 12, 1873. In 1915 he was elected to the Hall of Fame for Great Americans. A boulder from the Aar glacial moraine marks his grave at Mount Auburn, Cambridge.

Jean de Charpentier was born on December 8, 1786, in Freiberg, Saxony, Germany. His father was a professor at the Mining Academy, where Jean also studied under Abraham Gottlob Werner (1749–1817). He then worked as an engineer in the mines of Silesia before, in 1813, being appointed director of the salt mines at Bex, Switzerland.

In 1818 an ice dam broke causing a catastrophe in which many people died. The event affected Charpentier deeply, and he began extensive field studies of the geology of the Alps. In the course of these he became interested in large *erratic* boulders distributed across low ground. He rejected the theory that they

Louis Agassiz (1807–73), the Swiss-born American zoologist and glaciologist, who found convincing evidence of an ice age that once gripped the Northern Hemisphere. The picture is from a photograph of Agassiz taken in about 1850, when he was 43. *(Hulton Archive/Getty Images)*

were meteorites on the grounds that they were similar to rocks found in the Alps and did not believe they had been transported by floods, because he could not understand where so much water could have originated or disappeared to. He developed a theory of glacial transportation, which he first presented in 1835. It aroused little interest but did attract the attention of Louis Agassiz. Charpentier published his theory in 1841 as *Essai sur les glaciers* (Essay on glaciers), but it was overshadowed by Agassiz's work published a year earlier. Charpentier died at Bex on December 12, 1855.

THE CAUSE OF ICE AGES

Once scientists accepted that ice ages had occurred they began to speculate about what caused them. Several theories were advanced. Some scientists suggested that the amount of energy radiated by the Sun changes from time to time and that when it decreases temperatures on Earth fall. Others thought it possible that in the course of its travels through the Milky Way, the solar system periodically passes through clouds of dust and gas, which screen out some of the solar radiation reaching Earth. Both of these events are known to occur, and smaller climate changes are certainly linked to changes in the Sun, albeit indirectly, but neither is strong enough to trigger the onset or ending of an ice age.

The first hint at the explanation scientists accept today appeared in 1842, when the French mathematician Joseph-Alphonse Adhémar (1797–1862) published a book entitled *Les Revolutions de la Mer* (Revolutions of the sea). In his book Adhémar suggested an astronomical cause for ice ages. The Earth moves around the Sun in an elliptical orbit with the Sun at one focus of the ellipse. Consequently, the distance between the Earth and Sun changes through the year. *Perihelion*—when the Earth is closest to the Sun—occurs on December 22–23, and *aphelion*—when the Earth is farthest from the Sun—occurs on June 21–22. The following illustration, with the elliptical shape of Earth's orbit exaggerated, shows Earth's position at perihelion. Adhémar noted two consequences of the shape of Earth's orbit. The first is that Earth is at perihelion during the Northern Hemisphere winter, which means the Northern Hemisphere receives more solar radiation during its winter than the Southern Hemisphere does during its winter, when Earth is at aphelion. The second

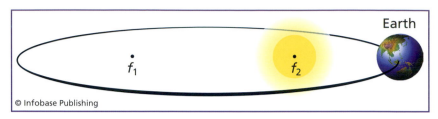

© Infobase Publishing

Precession of the equinoxes. The Earth moves around the Sun in an elliptical orbit, with the Sun at F_2, one of the two foci (F_1 and F_2). At present Earth is closest to the Sun on December 22–23.

consequence results from Kepler's second law of planetary motion, formulated in about 1610 by Johannes Kepler (1571–1630). This states that an imaginary line connecting Earth (or any other planet) to the Sun sweeps out equal areas in equal times throughout the body's orbit. This means that Earth moves faster as it approaches perihelion and slows around aphelion, so that Earth takes approximately seven days longer to travel past its aphelion than it does past its perihelion, making winter in the Southern Hemisphere about seven days longer than winter in the Northern Hemisphere.

At present, therefore, winters are longer in the Southern Hemisphere. Adhémar knew, however, that the dates of the equinoxes change over a cycle of about 22,000 years, a phenomenon known as the *precession of the equinoxes.* He reasoned that the present dates of aphelion and perihelion were producing an ice age in high latitudes of the Southern Hemisphere—Antarctica—and the Northern Hemisphere would experience an ice age when Earth is at aphelion in December. Ingenious though it was, Adhémar's theory was insufficient to produce an effect big enough to trigger an ice age and his calculation of the rate of the precession of the equinoxes was incorrect; the true period is 25,800 years (usually rounded to 26,000 years). Critics also pointed out that while the relative proportions of solar radiation reaching each hemisphere change due to precession, the total amount of radiation reaching Earth remains constant.

In 1864 the Scottish geologist and climatologist James Croll (1821–90) added a second component to Adhémar's theory. He proposed that ice ages occur approximately every 100,000 years and that they are triggered by the precession of the equinoxes combined with changes in the *eccentricity* of the Earth's orbit—the extent to which the orbit becomes more or less elliptical. This does change over a cycle of about 100,000 years, due to the gravitational influences of Jupiter and Saturn, and when the orbit is at its most elliptical Earth travels farther from the Sun and, therefore, it receives less solar radiation. Croll maintained that extreme eccentricity triggers an ice age in one or other hemisphere depending on the orbital location of Earth at the equinoxes. Croll also took account of the *albedo* effect—the reflectivity of surfaces. Ice and snow reflect most of the solar radiation falling upon them, so as the area of ice and snow increases more and more radiation is reflected back into space, causing surface temperatures to fall further.

James Croll was born near Cargill, Perthshire, on January 2, 1821. The family was poor, and James left school at 13. He read avidly, but worked at a succession of jobs in order to earn a living until, in 1859, he was appointed the keeper—in effect the janitor—of the Andersonian College and Museum (now the University of Strathclyde) in Glasgow. The museum housed a library, which James was permitted to use, and he taught himself physics, chemistry, geology, and astronomy. In 1867 he was placed in charge of the Edinburgh office of the Geological Survey of Scotland, a position he held until he retired in 1880, due to ill health. He was elected a fellow of the Royal Society and received an honorary degree from the University of St. Andrews, both in 1875. He published a number of books, including *Climate and Time, in Their Geological Relations* (1875) and *Discussions on Climate and Cosmology* (1885). James Croll died near Perth on December 15, 1890.

The theory of ice ages was finally completed in 1920, when, through its Paris publisher Gauthier-Villards, the Serbian Academy of Sciences and Arts published a monograph entitled *Théorie mathématique des phénomènes thermiques produits par la radiation solaire* (Mathematical theory of thermal phenomena produced by solar radiation). This proposed that over the last few million years the onset and ending of ice ages has been caused by three astronomical cycles: the eccentricity of Earth's solar orbit; the tilt of Earth's rotational axis; and the precession of the equinoxes due to axial wobble. These are shown in the following illustration. The author of the monograph was a Serbian civil engineer and geophysicist Milutin Milankovitch (1879–1958). Milankovitch had calculated the periods of these cyclical changes in Earth's orbit and rotation and compared them with the occurrences of ice ages. Changes in orbital stretch (eccentricity) alter the amount of radiation Earth receives from the Sun. Changes in the axial tilt, between 21.5° and 24.5° (at present it is 23.44° and decreasing), affect the intensity of solar radiation at high latitudes—the greater the tilt angle the stronger the sunlight at high latitudes. Axial wobble causes the precession of the equinoxes. When these three cycles coincide their effects are sufficient to trigger or end an ice age. There were many objections to the idea, but in 1924 Wladimir Köppen (see "Wladimir Köppen and His Classification" on pages 168–171)

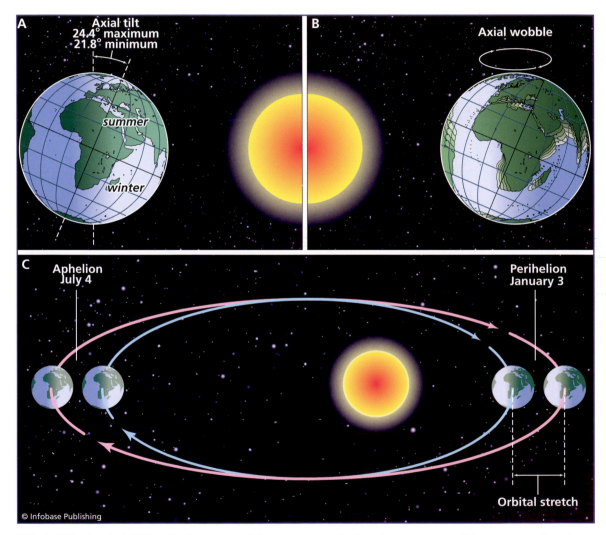

Milutin Milankovitch (1879–1958) proposed three astronomical cycles as causes of the onset and end of ice ages. These are, with their periods: orbital stretch (eccentricity) 100,000 years; axial tilt (obliquity) 41,000 years; and axial wobble (precession of the equinoxes) 26,000 years.

and Alfred Wegener (1880–1930) included it in their influential book *Climates of the Geological Past.* Finally, evidence of past climates and dates obtained from cores taken from seabed sediments led in 1976 to scientific acceptance of the theory.

Milutin Milankovitch was born on May 28, 1879, at Dalj (or Dali) near Osijek in what was then Austria-Hungary and is now Croatia. He studied at the Technische Hochschule in Vienna (now the Vienna University of Technology), graduating in civil engineering in 1902 and earning a doctorate in technical sciences in 1904. He worked as a civil engineer for the Viennese firm of Adolf Baron Pittel Betonbau-Unternehmung until 1909, when he obtained the position of professor of applied mathematics at the University of Belgrade. The Balkan Wars broke out in 1912 and were followed in 1914 by the start of World War I, when Milankovitch was interned by the Austro-Hungarian army. He was held for a time at Nezsider, but was then moved to Budapest, where he was allowed to work in the library of the Hungarian Academy of Sciences. He spent the remainder of the war studying solar climates and planetary temperatures. After the war he returned to his position at the University of Belgrade, where he remained for the rest of his life. Milankovitch died in Belgrade on December 12, 1958.

SUNSPOTS AND CLIMATE CYCLES

Many Christmas cards depict winter scenes of snow-covered landscapes, heavily muffled carol-singers, and bow-fronted shop windows in what are meant to be English village streets, with snow along the edges of the window frames. In fact, white Christmases are rare in Britain, and the popular image of cold, snowy Christmases is largely the invention of Charles Dickens (1812–70) in his stories, *A Christmas Carol* and *Pickwick Papers.* Dickens did not simply make up his descriptions, however. In his day English winters really were colder, and the winters of 1813 and 1814 were especially severe. Cold winters were common between about 1550 and 1850, a period known as the Little Ice Age, from which the world has only recently finished emerging. The Little Ice Age probably ended in about 1940, following a period of climatic warming that began in the middle of the 19th century. After 1940 global average temperatures fell very slightly until about 1976, after which they began rising once more.

Prior to the Little Ice Age the world was much warmer—and there is evidence that at times it was warmer than it is today. In those days England was a major producer of high-quality wine, with vineyards extending into the north of the country. Today England produces wine once more, but only in the south. Before the Little Ice Age, farmers grew wheat around Trondheim, Norway, and barley inside the Arctic Circle, and Norse settlers established farms in Greenland, farms that were abandoned early in the 13th century when increasing sea ice cut off their trading routes with Scandinavia. That time of benign climates comprised the Medieval Warm Period. Still earlier there was a Dark Ages Cool Period and before that a Roman Warm Period. Episodes of warm and cool climates alternate with a period of approximately 1,500 years. Today we appear to be entering what climate scientists are calling the Current Warm Period, though it is too soon to be certain whether it will continue, and many climate scientists believe human activity is affecting the global climate.

In 1889 Friedrich Wilhelm Gustav Spörer (1822–95), a German solar astronomer, noticed that between 1400 and 1510, a period when temperatures were especially low, very few sunspots were recorded. Spörer began his solar observations in 1858 and in 1874 he joined the staff of the Potsdam Astrophysical Laboratory, becoming its chief observer in 1882. He was experienced and greatly respected among solar astronomers, and the period of low sunspot activity he identified has become known as the Spörer Minimum. Spörer also discovered another sunspot minimum, in the 17th century.

Sunspots have interested astronomers for thousands of years. Chinese astronomers made the first written records of them in about 800 B.C.E. Sunspots are dark patches on the visible surface of the Sun, marking places where the temperature is about 2,700°F (1,500°C) cooler than the surrounding area. They are caused by intense magnetic fields, and most of the time their number increases and decreases over a cycle of approximately 11 years. During each cycle the sunspots change the latitude in which they first appear, and sunspots in different latitudes move across the solar surface at different rates.

Edward Walter Maunder (1851–1928), an English solar astronomer, was intrigued by Spörer's discovery. Maunder was born in London on April 12, 1851, and educated at King's College, London (now part of the University of London). After graduating he obtained a post at a bank, and, in 1873, after passing the necessary civil service examination, he was appointed as a photographic and spectroscopic assistant at the Royal Observatory, Greenwich, where his job was to photograph sunspots and measure their areas and positions. In 1891 Annie Scott Dill Russell (1868–1947), a brilliant mathematician, came to work at the observatory as a "lady computer." A "computer" was a person who performed mathematical computations. She and Maunder married and became professional colleagues. In 1873 Maunder was made a fellow of the Royal Astronomical Society. He died at Greenwich on March 21, 1928.

When Maunder read Spörer's paper he began searching the Royal Observatory's historical records to check whether Spörer was correct and whether there were other periods with few sunspots. His search identified a period from 1645 to 1715 during which very few sunspots were recorded and within that period there were 32 years when not a single sunspot was seen. In 1716, as the sunspot cycle was starting again, the English Astronomer Royal Edmond Halley (see "Edmond Halley, George Hadley, and the Trade Winds" on pages 115–118) wrote a paper about what he said was the first sunspot he had ever seen. In 1976 the American solar astronomer Jack (John A.) Eddy (born 1932) named the period from 1645 to 1715 the Maunder Minimum.

Since then other sunspot minima have been identified. John Dalton (see "John Dalton and Water Vapor" on pages 26–34) discovered one lasting from 1790 to 1820, the Dalton Minimum. The Wolf Minimum from 1280 to 1340 was discovered by the Swiss astronomer Johann Rudolf Wolf (1816–93) and the Dutch astrophysicist Jan Hendrik Oort (1900–92) discovered the Oort Minimum, from 1010 to 1050. There have also been periods of high sunspot numbers. One occurred between 1100 and 1250, coinciding with the Medieval Warm Period and another began in 1950. Such sunspot maxima seem to coincide with periods of warm climate and sunspot minima with periods of cool climate. This could be a coincidence, but the correlation appears to be very close. In 1997 Henrik Svensmark, director of the Center for Sun-Climate Research at the Danish Space Research Institute, and Eigil Friis-Christensen, director of the Danish National Space Center, proposed an explanation for the link. During sunspot maxima the *solar wind*—a stream of charged particles emitted by the Sun—intensifies, deflecting the cosmic radiation reaching Earth from outside the solar system. Cosmic radiation striking air molecules causes particles to form onto which water vapor can condense. Consequently, cosmic radiation plays a part in cloud formation. During sunspot minima the solar wind is weak and more cosmic radiation is able to penetrate the atmosphere, causing an increase in cloudiness that shades the surface, resulting in a fall in temperatures. When sunspot activity is high, less cosmic radiation can penetrate, so fewer clouds form and temperatures rise. Svensmark has observational and experimental evidence to support this theory, which contradicts the more widely accepted proposition linking climate change to the greenhouse effect (see sidebar "The Greenhouse Effect" on page 215).

GLOBAL WARMING

In 1827 the French mathematician and physicist Jean-Baptiste-Joseph Fourier (1768–1830) published an article in the *Mémoires de l'Académie Royale des Sciences de l'Institut de France* (Memoirs of the Royal Academy of Sciences of the Institute of France) in which he discussed the temperatures of the Earth and other planets. In it Fourier aimed to establish the study of planetary temperatures as a proper topic for physicists to consider. He explained how the warmth that the Earth receives from the Sun is distributed through the movements of air and water, and how all planets lose heat by radiation at infrared wavelengths—he called this *chaleur obscure* (dark heat)—thereby maintaining a thermal balance. He referred to the experiment Saussure conducted to measure the intensity of solar radiation, but drew an entirely different lesson from it (see "Horace-Bénédict de Saussure, the Hair Hygrometer, and the Weather House" on pages 77–81). Fourier observed that the glass of Saussure's box and also the Earth's atmosphere are transparent to sunlight but partially opaque to infrared radiation. Fourier is thus credited with discovering the greenhouse effect (see sidebar below), although he wrote of a hothouse rather than a greenhouse.

In the years that followed Fourier's paper other physicists became interested in the Earth's energy balance, one of the most influential being John Tyndall (1820–93). Starting in about 1859, Tyndall found that while oxygen, nitrogen, and hydrogen are completely transparent to infrared radiation other gases, especially water vapor, carbon dioxide, and ozone, are opaque. He had identified the atmospheric gases that caused the opacity Fourier had observed. In 1869 Tyndall

THE GREENHOUSE EFFECT

A body that is warmer than its surroundings loses energy by emitting electromagnetic radiation—radiation ranging from gamma and X-rays, through visible light and infrared (radiant heat) to microwaves and radio waves—and it continues to do so until it is at the same temperature as its surroundings. The greater the difference in temperature between the body and its surroundings the greater is the intensity of its radiation. Radiation travels at the speed of light. Consequently, its speed is constant regardless of its intensity, and its intensity can increase only by reducing the wavelength. The hotter the body is, the shorter is the wavelength of its radiation. The Sun is hot and radiates most intensely at short wavelengths. Its radiation warms the Earth's surface, causing Earth to radiate, but at much longer wavelengths because it remains relatively cool. Earth emits radiation at wavelengths between 5 μm and 50 μm, with a strong peak at 12 μm. (One micrometer (μm) is equal to 0.000001 meter or approximately 0.00004 inch.)

Air consists mainly of nitrogen (78.08 percent) and oxygen (20.95 percent). These gases are transparent to radiation at all wavelengths. The air also contains up to about 4 percent of water vapor, which is opaque to radiation at 5.3–7.7 μm and to all wavelengths above 20 μm. Except where the air is dry and contains almost no water vapor, water vapor absorbs much of the radiation leaving Earth's surface at those wavelengths and water droplets in clouds absorb more. The absorption of radiation warms the water vapor and air, but some distance above the surface the warm water vapor, rising by convection, radiates into space the energy that it absorbed lower down. The result is that warmth leaving Earth's surface warms the atmosphere as it rises and Earth's loss of warmth is delayed. This is the greenhouse effect.

The name "greenhouse effect" is misleading. In a real greenhouse sunlight warms the interior surfaces and the air inside the greenhouse warms by contact with those surfaces. The glass (or plastic) of the greenhouse prevents the warm air from rising by convection and cooling. Outside the greenhouse, however, warm air rises and cools, thereby helping Earth lose the solar warmth it has absorbed.

Although water vapor and clouds account for about 95 percent of the total greenhouse effect, other gases also contribute. Carbon dioxide absorbs radiation at 13.1–16.9 μm and accounts for about 3.6 percent of the total greenhouse effect, or 72.4 percent if the contribution from water vapor is omitted. Methane absorbs radiation at about 3.5 μm and 8 μm and nitrous oxide (N_2O) absorbs principally at about 5 μm and 8 μm. No atmospheric gas absorbs radiation at wavelengths between 8.5 μm and 13.0 μm, this waveband comprising an atmospheric window through which radiation escapes freely.

If there were no greenhouse effect, but there were clouds to reflect 31 percent of solar radiation, the average surface temperature would be about 5°F (-15°C). If there were no clouds the average temperature would be about 30°F (-1°C). The amount of water vapor in the atmosphere varies greatly from place to place and from time to time, so its contribution to the greenhouse effect is greater in some places and at some *(continues)*

(continued)
times than at others, but it remains constant over all.

In recent times the burning of carbon-based fossil fuels—coal, peat, natural gas, and petroleum—has released carbon dioxide into the atmosphere, increasing its atmospheric concentration since preindustrial times from about 0.028 percent to about 0.038 percent, and concentrations of other so-called greenhouse gases have also increased due to industrial activity, vehicle emissions, and agriculture. These now contribute to the natural greenhouse effect, producing an enhanced greenhouse effect. Their absorption of radiation is likely causing the atmosphere to become warmer, although the extent of this effect is uncertain and controversial.

also discovered that light passes unimpeded through a solution or pure solvent, but that a light beam is visible in a colloidal solution—a *colloid* consists of particles larger than molecules but too small to be visible in an ordinary microscope evenly dispersed throughout another substance. He suggested that atmospheric molecules behave like colloidal particles, scattering light and scattering blue light most strongly. This, he said, would make the sky blue, a suggestion Lord Rayleigh (1842–1919) confirmed in 1871 (see "Lord Rayleigh, William Ramsay, Noble Gases, and Why the Sky Is Blue" on pages 36–41).

John Tyndall was born on August 2, 1820, at Leighlin Bridge, County Carlow, Ireland, the son of a police officer. He attended school in the nearby town of Carlow, and when he was 17 he joined the Ordnance Survey—the British agency responsible for mapping the country—to train as a surveyor. He left the Ordnance Survey in 1843 and worked for a time as a surveyor for the railroad industry. In 1847, when the railroad boom ended, he obtained a teaching post and in 1848 he enrolled at the University of Marburg, Germany, where he studied physics, calculus, and chemistry. His chemistry teacher was Robert Bunsen (1811–99), who helped and encouraged him. Tyndall gained a doctorate in 1850. He returned to his old teaching job in 1851, where he supplemented his salary by translating and reviewing scientific articles. By this time his reputation was growing, and in 1852 he was elected a fellow of the Royal Society. In 1853 he was invited to deliver a lecture at the Royal Institution. This was so successful that he was asked to deliver a series of lectures. He was later appointed professor of natural philosophy at the Royal Institution and succeeded Michael Faraday (1791–1867) as its superintendent. The photograph of John Tyndall was taken on January 1, 1884. He died on December 4, 1893, from an accidental overdose of chloral hydrate, which he was prescribed to help him sleep.

Until this time no one had attempted to measure the amount of radiation different gases absorb and the effect this might have on temperature. The first person to do so was a professor of physics at the Stockholms Högskola (high school, equivalent to the science faculty of a university), Svante August Arrhenius (1859–1927). In April 1896 Arrhenius published a paper, "On the Influence of Carbonic Acid in the Air upon the Temperature of the Ground," in *The London, Edinburgh, and Dublin Philosophical Magazine and Journal of Science.* Arrhenius began by paying tribute to John Tyndall and other scientists and then introduced his calculations. He had calculated the change in temperature that would result if the atmospheric carbon dioxide (carbonic acid) concentration were to be 67 percent, 150 percent, 200 percent, 250 percent, and 300 percent of its 1896 value. He performed the calculation for 13 latitudinal belts, each of 10 degrees, from 70°N to 60°S, with the temperatures for each of the four seasons and the annual average for each belt. It was an astonishing achievement, for Arrhenius performed all the calculations with only a slide rule to help him. He found that doubling the concentration of carbon dioxide would increase the mean annual temperature at the equator by 8.91°F (4.95°C) and at 60°N by 10.89°F (6.05°C).

John Tyndall (1820–93) was an Irish physicist who studied how certain gases absorb infrared radiation. *(General Photographic Agency/Getty Images)*

Each year the Asahi Glass Foundation awards its Blue Planet Prizes for contributions to environmental understanding and improvement. In 1998 the recipient of one Blue Planet Prize was Mikhail Ivanovich Budyko (1920–2001), a Belorussian climatologist who altered the perspective from which scientists viewed climates. Previously climatologists had worked by studying data gathered from all over the world. Budyko studied the radiation balance of the Earth, changing climatology into a much more quantified, physical discipline. In 1956 he published his findings in a book entitled *Heat Balance of the Earth's Surface,* describing how he had studied the amounts of incoming and outgoing energy in several parts of the world, verified his calculations against measurements made in those places, and then applied the

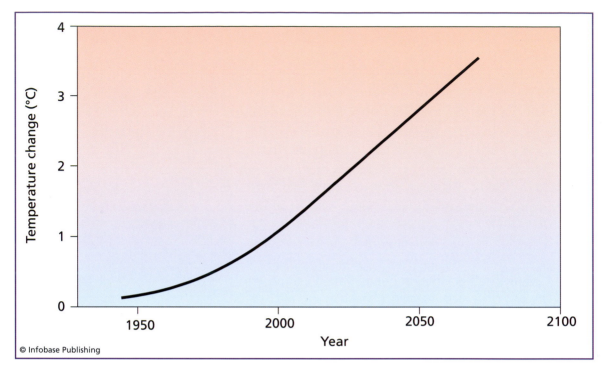

In 1972 Mikhail I. Budyko (1920–2001) calculated the effect on the global atmospheric temperature of the continuing increase in the concentration of carbon dioxide. This is the graph he produced. It was accepted in 1982 by a meeting of scientists from the United States and Soviet Union.

method he had developed to weather data from all over the world to calculate heat balances everywhere. In 1963 Professor Budyko published an atlas that showed the energy balances of every part of the Earth. His atlas was the starting point from which other climate scientists developed their own research.

In 1972 Budyko challenged the prevailing view among climatologists that the global climate was entering a cooler phase. He maintained that the burning of fossil fuels was increasing the atmospheric concentration of carbon dioxide, and that this would lead to a rise in average temperatures. The illustration shows the graph of his calculation that the atmosphere might warm by about 6.3°F (3.5°C) by 2075. Mikhail Budyko inspired other climate scientists, and today the Intergovernmental Panel on Climate Change assembles and disseminates the work of his successors. Many climate scientists now believe that the accumulation of greenhouse gases will lead to rising temperatures in years to come. Most governments now accept this theory, although a significant number of scientists disagree, maintaining that the calculations on which the theory is based greatly exaggerate the extent and consequences of potential greenhouse warming.

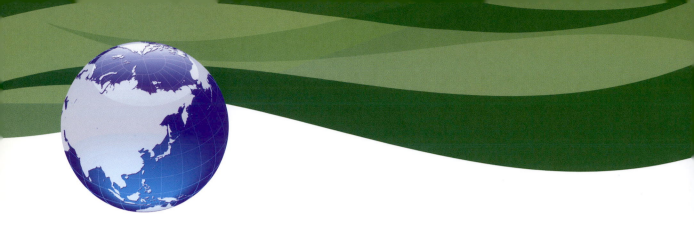

Conclusion

Satellites orbiting the Earth constantly monitor the formation, movement, and dissolution of weather systems. Their instruments measure changes in the height of sea level, the temperature of the air and sea surface, the advance and retreat of ice and snow, and changing patterns of vegetation. A suite of 3,000 Argo floats, distributed throughout the world's oceans, drifts 6,500 feet (2,000 m) below the ocean surface automatically measuring the temperature and salinity of the water and the speed and track of ocean currents. Every 10 days each float rises to the surface, transmits its data and position to a satellite that relays it to a surface station, and sinks once more to the depths.

Atmospheric scientists have a vast array of tools for gathering data and powerful supercomputers with which to process them, but all of this technology is new. The first weather satellite, TIROS (Television and Infrared Observation Satellite), was launched by the United States on April 1, 1960. Today meteorologists rely on supercomputers, and climate modeling would be impossible without them. The first supercomputers were built in the 1960s. The last of the Argo floats was deployed in October 2007. It is less than 170 years since the first telegraph message was sent the 40 miles (64 km) from Washington, D.C., to Baltimore, providing a way to gather data from a wide area in time to issue weather reports and, a little later, weather forecasts.

It is not surprising, therefore, that scientists still have much to discover about the Earth's atmosphere and its behavior. Although people have been studying the air and the weather for thousands of years, and the earliest book on weather was written more than 2,000 years ago, only recently has it become possible to observe the entire atmosphere as a whole, measuring changes as they occur.

This book has used a series of snapshots—short accounts of discoveries and the scientists who made them—to tell how knowledge of the atmosphere was acquired over the centuries. Inevitably, the picture is incomplete. But perhaps it is sufficient to give an impression of the curiosity that has driven investigators and their excitement at finding answers to questions that would not let them rest. The part of the story told here began in ancient Greece and ends with the start of modern concerns about climate change.

It is not the end, of course. Atmospheric research is advancing rapidly. In years to come everyone may be able to access at any time of day or night a highly accurate personal weather forecast for a particular place for at least a week ahead. Warning systems, and the ability to respond appropriately to them, may render extreme weather harmless. And one day it may be possible to control weather.

GLOSSARY

advection fog fog that forms in warm, moist air if it moves across a cold surface.

aerosol a mixture of small solid and liquid particles that remains suspended in the air.

air mass a large body of air, extending over much of a continent or ocean, throughout which the physical characteristics are approximately uniform at every height.

Aitken nucleus an airborne particle that is less than 0.0016 inch (0.4 μm) across.

albedo the reflectivity of a surface to solar radiation.

Amontons's law for any given change in temperature, the pressure exerted by a gas always changes by the same amount; $P_1T_2 = P_2T_1$, where P_1 and P_2 are the initial and final pressures and T_1 and T_2 are the initial and final temperatures.

anemometer an instrument that measures wind speed.

aneroid barometer a **barometer** that uses no liquid.

angular velocity (Ω) speed of movement along a curved path, usually expressed in radians per second (rad/s); the circumference of a circle is equal to 2π radians, therefore, $Ω = 2π/T$ rad/s, where T is the time in seconds taken to complete one revolution.

anticyclone a center of high pressure around which air circulates anticyclonically (clockwise in the Northern Hemisphere).

aphelion the point in its orbit where Earth is farthest from the Sun.

barometer an instrument that measures atmospheric pressure.

boundary layer the very thin layer of air that is in direct contact with a surface; see also **planetary boundary layer**.

Boyle's law the volume occupied by a gas is inversely proportional to the pressure under which the gas is held.

butterfly effect a metaphor that illustrates an extreme sensitivity to initial conditions displayed by certain systems, including weather systems, and summarized as the suggestion that a butterfly flapping its wings in Brazil might trigger a tornado in Texas; it means that differences so small as to be undetectable can cause apparently similar weather systems to develop in radically different ways, making long-term prediction impossible.

CCN See **cloud condensation nuclei**.

Charles's law for any given change in temperature the volume of a gas always changes by the same amount; $V_1/T_1 = V_2/T_2$, where V_1 and T_1 are the initial volume and temperature and V_2 and T_2 are the final volume and temperature.

clepsydra a clock operated by moving water.

climate the average of the weather conditions experienced in a particular place over a long period, commonly 30 years.

cloud condensation nuclei (CCN) airborne particles onto which water vapor condenses to form clouds.

cold cloud a cloud that contains ice crystals and **supercooled** water droplets.

cold front a **front** behind which the air is cooler.

colloid a system comprising particles larger than molecules but too small to be visible in an ordinary microscope that are dispersed evenly throughout another substance.

condensation the change of phase from gas to liquid.

conduction the transfer of heat through direct contact between bodies at different temperatures.

conservation of mass, law of the mass contained within a closed system will remain constant regardless of the processes occurring inside the system.

convection the transfer of heat by the movement of a gas or liquid.

convergence the flow of air from all directions toward a central point.

CorF See **Coriolis effect**.

Coriolis effect (CorF) the apparent deflection, to the right in the Northern Hemisphere and to the left in the Southern Hemisphere, experienced by bodies moving long distances over the Earth's surface, but remaining unattached to the surface.

cyclone a center of low pressure around which air circulates cyclonically (counterclockwise in the Northern Hemisphere).

Dalton's law of partial pressures a law proposed in 1803 by John Dalton (1766–1844) stating that the pressure exerted by a mixture of gases is equal to the sum of the **partial pressure** exerted by each gas separately.

deposition   the direct change of phase in water from gas to solid, without passing through the liquid phase.

depression   a center of low pressure at the crest of a frontal wave.

dew-point depression   the difference between the ambient air temperature and the **dew-point temperature**.

dew-point temperature   the temperature at which atmospheric water vapor condenses.

divergence   the outward flow of air in all directions from a central point.

dry-bulb thermometer   a thermometer with its bulb exposed directly to the air.

eccentricity   the extent to which a planetary orbit deviates from being circular; Earth's solar orbit varies over a cycle of 100,000 years from being almost circular to being slightly elliptical.

elements   See **principles**.

El Niño   a weakening or in extreme cases reversal of the trade winds and South Equatorial Current in the equatorial South Pacific Ocean that occurs at intervals of 2 to 7 years and produces widespread climatic effects, including heavy rain in western South America and drought in Indonesia; the opposite change is called **La Niña**.

ENSO   abbreviation for an **El Niño-Southern Oscillation** event.

equinox   one of the two days each year, on March 20–21 and September 22–23, when at noon the Sun is directly overhead at the equator, and the Sun remains above the horizon for 12 hours and below it for 12 hours everywhere in the world.

erratic   a rock or mass of smaller material such as gravel that is of a type different from the underlying bedrock and that has been glacially transported to the location where it is found.

evaporation   the change of phase from liquid to gas.

evaporimeter   an instrument that measures the rate of **evaporation** of water.

exosphere   the outermost region of the atmosphere, extending from the **thermopause** but with no defined upper boundary.

fiducial point   in calibrating an instrument, a fixed value (e.g., the freezing and boiling temperatures of water) to which other values can be related.

freezing   the change of phase in water from liquid to solid (ice).

front   a boundary between two **air masses**.

gas laws   a set of formal descriptions of the relationships between the temperature, volume, and pressure of an ideal gas.

glacier   a large mass of ice resting wholly or principally on solid ground and typically moving slowly.

heat island   an urban area that is warmer than the surrounding countryside, due to the release of warm air from buildings, the absorption of heat by dark surfaces, and the reduction in wind speed caused by friction with buildings.

heterosphere   the region of the upper atmosphere in which the chemical composition of the air changes with increasing height.

homosphere   the region of the atmosphere, including the **troposphere, stratosphere,** and lower **mesosphere,** within which the chemical composition remains constant.

humidity   a measure of the amount of water vapor present in the air.

hydrogen bond   a chemical bond linking the hydrogen (H^+) atom of one **polar molecule** with an electronegative atom in another molecule of the same compound (e.g., between the H^+ of one water molecule and the O^- of another).

hygrometer   an instrument for measuring atmospheric **humidity**.

ion   an atom or molecule that has gained or lost one or more electrons and, consequently, carries a positive or negative electromagnetic charge.

isobar   a line drawn on a map to link places where the atmospheric pressure is the same.

La Niña   a strengthening of the trade winds and South Equatorial Current in the equatorial South Pacific Ocean that occurs at intervals of 2 to 7 years and produces widespread climatic effects, including heavy rain in Indonesia and drought in western South America; the opposite change is called **El Niño**.

latent heat   the energy that is absorbed or released when a substance changes phase between solid, liquid, and gas; energy is absorbed when the substance moves to a higher energy state (solid to liquid and liquid to gas) and released when it moves to a lower energy state (gas condenses, liquid solidifies).

lower atmosphere   the **troposphere**.

melting   the change of phase in water from solid (ice) to liquid.

mesopause   the upper boundary of the **mesosphere** at an average height of 50 miles (80 km).

mesosphere   the region of the atmosphere that extends from the **stratopause** to the **mesopause**.

occlusion   a situation where a **cold front** advancing against a **warm front** raises the warm air clear of the ground.

partial pressure   the proportion of the pressure exerted by a mixture of gases that can be ascribed to a single ingredient of that mixture.

perihelion   the point in its orbit where Earth is closest to the Sun.

phlogiston   a substance once believed to reside in all combustible substances and that was released into the air when the substance was burned.

photolytic   light-driven.

planetary boundary layer   the lowest layer of the atmosphere, extending from the surface to an altitude of about 1,500 feet (460 m), in which the characteristics of the air are significantly influenced by the proximity of the surface and winds are affected by friction with the surface.

polar front   the **front** between tropical and polar air.

polar molecule   a molecule that is electromagnetically neutral over all, but carries a small positive charge on one side and a negative charge on the other; water is a polar molecule.

polygon   a plane (two-dimensional) shape that has three or more angles and as many straight sides as it has angles (e.g., triangle, square).

polyhedron   a three-dimensional solid constructed from faces each of which is a plane **polygon**.

precession of the equinoxes   the change in the dates when Earth is at **aphelion** and **perihelion** over a cycle of 25,800 years.

principles (elements)   earth, water, air, and fire, the four basic forces or ideas that, in the opinion of Greek philosophers, regulated everything that moves.

psychrometer   an instrument that measures **humidity** by comparing readings from a **wet-bulb thermometer** with those from a **dry-bulb thermometer**.

pyrometer   an instrument that measures the temperature of an object from the radiation it emits.

Pythagorean Brotherhood   a Greek school of philosophy, founded in about 500 B.C.E. by Pythagoras (ca. 569–ca. 475 B.C.E.), that taught that the Earth, Sun, planets, and Moon are fixed to crystal spheres and that everything in the universe is formed from whole numbers.

radiation   the transfer of energy as electromagnetic rays (gamma, X-ray, light, heat, and radio waves) or a stream of particles.

radiosonde   a package of instruments carried beneath a weather balloon that transmits its data to a ground receiving station.

rainout the removal from the air of particles that act as **cloud condensation nuclei**.

regelation the process in which ice melts under pressure and then refreezes.

saturation vapor pressure the **vapor pressure** at which water vapor saturates the layer of air immediately adjacent to a surface at a given temperature.

scattering the repeated change of direction experienced by sunlight as it encounters atmospheric gas molecules and **aerosol** particles.

solar wind a stream of charged particles emitted by the Sun.

solstice one of the two days each year when at noon the Sun is directly overhead at the tropic of Cancer (June 20–21) or tropic of Capricorn (December 22–23); each solstice is midwinter day in one hemisphere, when hours of daylight reach a minimum, and midsummer day in the other hemisphere, when hours of daylight reach a maximum.

source area the region of the Earth where an **air mass** forms.

Southern Oscillation a change in the distribution of atmospheric pressure over the eastern Indian Ocean and South Pacific Ocean that occurs every 1 to 7 years and produces the cycle of **El Niño** and **La Niña** changes in the trade winds and South Equatorial Current.

storm glass a device for predicting the weather that was popular in the 18th and 19th centuries, but is no longer used.

stratopause the upper boundary of the **stratosphere**, at an average height of 30–37 miles (50–60 km).

stratosphere the region of the atmosphere that extends from the **tropopause** to the **stratopause**.

sublimation the direct change of phase in water from solid (ice) to vapor, without passing through the liquid phase.

supercooled describes water that remains liquid despite being below freezing temperature.

teleconnections linked atmospheric events that occur in widely separated regions of the world.

thermocouple a device consisting of two rods or wires of metals with different physical properties that are shaped into half loops and joined at their ends to form a circle; changes in temperature cause an electric current to flow around the circle.

thermopause the upper boundary of the **thermosphere,** at a height of 310–620 miles (500–1,000 km).

thermoscope an instrument that shows changes in temperature by a visual effect, but that has no scale against which the magnitude of change can be measured.

thermosphere the region of the atmosphere that extends from the **mesopause** to the **thermopause**.

Torricelli vacuum a vacuum produced by filling with mercury a tube with one open end, immersing the open end in a bath of mercury, and allowing mercury to flow from the tube while keeping the open end submerged; the method was first used in 1643 by Evangelista Torricelli (1608–47).

torsion balance a device consisting of a horizontal bar suspended from a thin fiber that acts as a very weak spring; if weights are suspended from the ends of the bar the fiber will twist, causing the bar to rotate, and the amount of rotation is proportional to the force applied to the bar.

trade winds the prevailing winds in the Tropics, blowing from the northeast in the Northern Hemisphere and from the southeast in the Southern Hemisphere.

tropopause the upper boundary of the **troposphere,** at an average height of 10 miles (16 km) over the equator, seven miles (11 km) in middle latitudes, and five miles (8 km) over the poles.

troposphere the lowest layer of the atmosphere, extending from the surface to the **tropopause**.

upper atmosphere all of the atmosphere above the **tropopause**.

valley glacier a **glacier** that flows between walls of rock.

vapor pressure the **partial pressure** exerted by atmospheric water vapor.

Walker cell a flow of air over the Tropics in which air rises, moves east or west at high level, and subsides, feeding air into the **trade winds**.

Walker circulation the overall pattern of **Walker cells**.

warm cloud a cloud that contains only liquid water droplets.

warm front a **front** behind which the air is relatively warmer.

washout the removal from the air of particles that collide with and adhere to falling raindrops.

wet-bulb thermometer a thermometer with the bulb wrapped in a muslin wick to keep it permanently wet.

wind rose a diagram showing the frequency with which the wind blows from each direction at a particular location.

FURTHER RESOURCES

Allaby, Michael. *Encyclopedia of Weather and Climate,* Rev. ed. 2 vols. New York: Facts On File, 2007. Short essays explaining the principles of meteorology and climatology.

———. *A Change in the Weather.* New York: Facts On File, 2004. An account of the history of climatic change.

———. *Droughts,* Rev. ed. New York: Facts On File, 2003. Includes an explanation of teleconnections and ENSO events.

———, and Derek Gjertsen. *Makers of Science.* 5 vols. New York: Oxford University Press, 2002. Brief biographies of scientists and descriptions of their achievements.

Hamblyn, Richard. *The Invention of Clouds: How an Amateur Meteorologist Forged the Language of the Skies.* New York: Farrar, Straus and Giroux, 2001. A biography of Luke Howard.

Intergovernmental Panel on Climate Change. *Climate Change 2007: The Physical Science Basis.* New York: Cambridge University Press, 2007. The technical explanation of the scientific background to the IPCC review of climate change.

Jardine, Lisa. *Ingenious Pursuits: Building the Scientific Revolution.* London: Little, Brown and Company, 1999. An account of the development of science in the 18th century by a leading historian.

Lamb, H. H. *Climate, History and the Modern World,* 2d ed. New York: Routledge, 1995. An account of historical climate change and its social, economic, and political consequences by one of the most respected climatologists of his generation.

Monmonier, Mark. *Air Apparent: How Meteorologists Learned to Map, Predict, and Dramatize Weather.* Chicago: University of Chicago Press, 1999. A history of the development of weather maps.

Newby, Eric. *The Last Grain Race.* London: Secker and Warburg, 1956. A personal account of life on board the sailing ship that won the last grain race.

Pain, Stephanie. "The accidental aeronaut," *New Scientist* (September 22, 2007): 54–55. The story of how a meteorologist inadvertently set a world altitude record.

Svensmark, Henrik, and Nigel Calder. *The Chilling Stars: A New Theory of Climate Change.* Thriplow, Cambridge, UK: Icon Books, 2007. A popular account of the possible link between solar activity and climate by a Danish climate physicist and the former editor of *New Scientist.*

WEB SITES

Adventures in Cybersound. "John Frederic Daniell: 1790–1845." Available online. URL: www.acmi.net.au/AIC/DANIELL_BIO.html. Accessed February 12, 2008. Biography of Daniell.

Aristotle. *Meteorology,* translated by E. W. Webster. The Internet Classics Archive by Daniel C. Stevenson. Web Atomics. Available online. URL: http://classics.mit.edu/Aristotle/meteorology.1.i.html. Accessed January 15, 2008. The text of Aristotle's *Meteorologica* in English.

Arrhenius, Svante. "On the Influence of Carbonic Acid in the Air upon the Temperature of the Ground." *The London, Edinburgh, and Dublin Philosophical Magazine and Journal of Science* (April 1869). Available online. URL: www.globalwarmingart.com/images/1/18/Arrhenius.pdf. Accessed April 3, 2008. The full text of Arrhenius's original paper.

Astronomical University, University of Uppsala. "Anders Celsius 1701–44." Available online. URL: www.astro.uu.se/history/Celsius_eng.html. Accessed February 7, 2008. Biographical and other information about Celsius.

Blamire, John. "John Dalton." *Science@a Distance.* Available online. URL: www.brooklyn.cuny.edu/bc/ahp/FonF/Dalton.html. Accessed January 24, 2008. Biographical information about Dalton.

Cavendish, Henry. "Experiments on Air." *Philosophical Transactions of the Royal Society* 75: 372. London: 1785. Available online. URL: http://web.lemoyne.edu/~giunta/cavendish.html. Accessed January 24, 2008. The original text of Cavendish's paper.

Chapman, Allan. "England's Leonardo: Robert Hooke (1635–1703) and the Art of Experiment in Restoration England." *Proceedings of the Royal Institution of Great Britain* 67: 239–275. Available online. URL: http://home.clara.net/rod.beavon/leonardo.htm. Accessed February 5, 2008. Description of Hooke's work and methods.

Eliassen, Arnt. "Jacob All Bonnevie Bjerknes." Washington, D.C.: National Academy of Sciences. Available online. URL: www.nap.edu/readingroom/books/biomems/jbjerknes.html. Accessed March 6, 2008. A biography of Bjerknes and account of his scientific work.

Ferrel, William. "An Essay on the Winds and Currents of the Ocean." *Nashville Journal of Medicine and Surgery,* vol. xi, nos. 4 and 5 (October and November 1856). Available online. URL: www.aos.princeton.edu/WWWPUBLIC/gkv/history/ferrel-nashville56.pdf. Accessed February 27, 2008. Full text of Ferrel's 1856 paper.

"Galton, Sir Francis, F.R.S. 1822–1911." Available online. URL: http://galton.org/. Accessed March 18, 2008. Home page of a Web site devoted to Galton, including a full biography.

Hochfelder, David. "Joseph Henry: Inventor of the Telegraph?" Washington, D.C.: Smithsonian Institution. Available online. URL: http://siarchives.si.edu/history/jhp/joseph20.htm. Accessed March 13, 2008. An account of Henry's contribution to the development of the telegraph.

Humphreys, W. J. "Giants of Science: Cleveland Abbe 1838–1916.) *NOAA History.* NOAA. Available online. URL: www.history.noaa.gov/giants/abbe.html#top. Accessed March 14, 2008. Genealogy and biography of Abbe.

Hunter, Michael. "The Life and Thought of Robert Boyle." London: Birkbeck College. Available online. URL: www.bbk.ac.uk/boyle/biog.html. Accessed February 20, 2008. An account of Boyle's scientific achievements.

Lavoisier, Antoine-Laurent. "Memoir on the Nature of the Principle Which Combines with Metals during Their Calcination and which Increases Their Weight." *Mémoires de l'Académie Royale des Sciences* (for 1775): 520–526. (Published Paris 1778). Available online. URL: http://web.lemoyne.edu/~giunta/lavoisier.html. Accessed January 21, 2008. Text of Lavoisier's original paper in English.

Loy, Jim. *Phlogiston Theory.* Available online. URL: www.jimloy.com/physics/phlogstn.htm. Accessed January 18, 2008. Explanation of the phlogiston theory.

Milikan, Frank Rives. "Joseph Henry: Father of Weather Service." Washington, D.C.: The Smithsonian Institution. Available online. URL: http://siarchives.si.edu/history/jhp/joseph03.htm. Accessed March 13, 2008. An account of Henry's establishment of a network of weather stations.

NOAA. "The William Ferrel Tide-Predicting Machine." *Tides and Currents.* NOAA. Available online. URL: http://tidesandcurrents.noaa.gov/predma1.html. Accessed February 28, 2008. Photographs of Ferrel's machine and information about it.

Nobel Prize in Chemistry 1904, biography. "Sir William Ramsay." Available online. URL: http://nobelprize.org/nobel_prizes/chemistry/laureates/1904/ramsay-bio.html. Accessed January 25, 2008. Biography of Ramsay.

Nobel Prize in Physics 1904, biography. "Lord Rayleigh." Available online. URL: http://nobelprize.org/nobel_prizes/physics/laureates/1904/strutt-bio.html. Accessed January 25, 2008. Biography of Lord Rayleigh.

O'Connor, J. J., and E. F. Robertson. "Blaise Pascal." St. Andrews: University of St. Andrews. Available online. URL: www-groups.dcs.st-and. ac.uk/~history/Biographies/Pascal.html. Accessed February 21, 2008. Biography of Pascal.

——. "Edmond Halley." St. Andrews: University of St. Andrews. Available online. URL: www-groups.dcs.st-and.ac.uk/~history/Biographies/Halley.html. Accessed February 26, 2008. Biography of Halley.

——. "Francis Galton." St. Andrews: University of St. Andrews. Available online. URL: www-groups.dcs.st-and.ac.uk/~history/Biographies/Galton.html. Accessed March 18, 2008. Biography of Galton.

——. "Lewis Fry Richardson." St. Andrews: University of St. Andrews. Available online. URL: www-history.mcs.st-andrews.ac.uk/Biographies/Richardson.html. Accessed March 18, 2008. Biography of Richardson.

——. "Robert Boyle." St. Andrews: University of St. Andrews. Available online. URL: www-groups.dcs.st-and.ac.uk/~history/Biographies/Boyle.html. Accessed February 20, 2008. A biography of Boyle.

——. "Vilhem Friman Koren Bjerknes." St. Andrews: University of St. Andrews. Available online. URL: www-history.mcs.st-andrews.ac.uk/Biographies/Bjerknes_Vilhelm.html. Accessed March 5, 2008. A biography of Bjerknes.

——. "William Ferrel." St. Andrews: University of St. Andrews. Available online. URL: www-groups.dcs.st-and.ac.uk/~history/Biographies/Ferrel.html. Accessed February 27, 2008. Biography of Ferrel.

Rayleigh, John William Strott, third baron. Nobel Prize in Physics 1904, biography. Available online. URL: http://nobelprize.org/nobel_prizes/physics. Accessed July 11, 2008.

Riebeck, Holli. "Paleoclimatology: Introduction." Earth Observatory, NASA. Available online. URL: http://earthobservatory.nasa.gov/Study/Paleoclimatology/. June 28, 2005. Accessed March 28, 2008. A description of the discovery of past ice ages.

Robert Hooke home page: A Web site devoted to Hooke. Available online. URL: www.roberthooke.com/robert_hooke_biography_001.htm. Accessed February 1, 2008. Information about Hooke's life and work.

Schultz, David M., and Robert Marc Friedman. "Tor Harold Percival Bergeron." In *New Dictionary of Scientific Biography* (ed. Noretta Koertge). Charles Scribner's Sons, 2007. Available online. URL: www.cimms.ou.edu/~schultz/papers/TorBergeron.pdf. Accessed February 14, 2008. Biography of Bergeron.

Sedlacek, Cheryl. "The Great Glacier Controversy." Available online. URL: www.emporia.edu/earthsci/student/sedlacek1/website.htm.

Accessed March 28, 2008. A brief account of the way scientists discovered that glaciers flow and were once more extensive than they are now.

Tibell, Gunnar. "Linnaeus' Thermometer." Uppsala: Uppsala Universitet. Available online. URL: www.linnaeus.uu.se/online/life/6_32.html. Accessed February 6, 2008. Description of the thermometer Linnaeus installed in the Uppsala Botanic Garden.

University of California Museum of Paleontology. "Aristotle (384–322 B.C.E.)." Available online. URL: www.ucmp.berkeley.edu/history/aristotle.html. Accessed January 15, 2008. Biography and other information about Aristotle.

Walker, J. M. "Pen Portrait of Sir Gilbert Walker, CSI, MA, ScD, FRS." *Weather,* vol. 52, no. 7, pp. 217–220. London: Royal Meteorological Society, 1997. Available online. URL: www.rmets.org/pdf/walkergt.pdf. Accessed March 6, 2008. A description and biography of Walker.

Wisniak, Jaime. "Guillaume Amontons." *Revista CENIC Químicas,* vol. 36, no. 3, 2005. Available online. URL: http://revistas.mes.edu.cu:9900/EDUNIV/03-Revistas-Cientificas/Rev.CENIC-Ciencias-Quimic as/2005/3/Q9-04.pdf. Accessed February 8, 2008. Biographical details about Amontons.

INDEX

Note: *Italic* page numbers indicate illustrations; page numbers followed by *m* indicate maps; page numbers followed by *t* indicate tables.